THE TECHNICAL ENTERPRISE

THE TECHNICAL ENTERPRISE
Present and Future Patterns

HERBERT I. FUSFELD

BALLINGER PUBLISHING COMPANY
Cambridge, Massachusetts
A Subsidiary of Harper & Row, Publishers, Inc.

224001

338.06
F 993

International Standard Book Number: 0-88730-033-2

Library of Congress Catalog Card Number: 86-3340

Printed in the United States of America

Library of Congress Cataloging-in-Publication Data

Fusfeld, Herbert I.
  The technical enterprise.

  Includes bibliographies and index.
  1. Technological innovations—Economic aspects.
2. High technology industries. I. Title.
HC79.T4F87 1986   338'.06   86-3340
ISBN 0-88730-033-2

CONTENTS

LIST OF FIGURES AND TABLES

PREFACE

Since its creation in 1950, the National Science Foundation has been the principal agency of the federal government charged with strengthening the nation's reservoir of basic science and engineering. Traditionally, NSF has carried out this mandate by funding university research. Academic researchers and administrators form the most consistent constituency of NSF, and look upon it as an important force for stability and growth of university research in a manner compatible with academic traditions.

In 1983 Roland Schmitt, senior vice president of General Electric Company for research and development, became chairman of the National Science Board, the governing body of NSF. He was only the second person with an industrial background to be in that position. Then in 1984, for the first time in the Foundation's history, an individual from industry was named as director of NSF. Erich Bloch had spent a long career at IBM in research, development, and management, and was vice president for technical personnel development when he joined NSF.

Now two individuals steeped in the structures and practices of industry, particularly industrial research, were responsible for setting the direction of that federal activity most closely linked to the health and growth of university research. Since both were, and are, good friends of mine, I received a number of calls from colleagues and from journalists curious about their probable approaches. Would two people from industry in those positions change the technical system, at least those parts affected by NSF?

I had two comments. First, any two intelligent and competent individuals would develop independent programs differing from those of any other two people regardless of background. Second, the questioners were confusing cause and effect.

The technical system was *not* going to change because two people from industry occupied the top positions at NSF and NSB. They were named *because* the technical system had changed. The Reagan administration did not create a new situation, as suggested by some of the callers. It simply recognized that the emperor had lost his clothes. The model of the technical system in the minds of many people which fit comfortably with a university researcher as head of NSF did not represent the process of technical change going into the twenty-first century. Our economic well-being and political security are affected directly by technical change, and the system that produces such change is itself in transition, reacting to pressures from the technical process and from our economic environment.

How does the technical system really operate? What has been changed and why? How will these changes affect and be affected by investment decisions, corporate strategies, government policies, university structures, and research management? That is, in great part, what this book is about.

ACKNOWLEDGMENTS

I am deeply indebted to all my friends within the technical enterprise who have been a source of knowledge and insight. A career in research management provided me opportunities for observation and analysis of the technical system at work. My colleagues and friends were, and are, the senior technical executives of major corporations in the United States and abroad, as well as leading government officials concerned with science policy throughout the OECD countries. My work and interests required involvement too with university researchers and administrators. The inputs and judgments of all these colleagues have provided important background for this book.

During my study and analyses about the technical system for this book, I have profited greatly from the discussions and working relations with Richard Nelson of Yale and Richard Langlois, University of Connecticut. Both have been active participants in the research programs of New York University's Center for Science and Technology Policy, and have kept me from stumbling too badly in the use of economic theories of technical change. Finally, this book would not have been possible without the competent and detailed research of Carmela Haklisch and Lois Peters. Their studies over the past seven years on different features of the technical system—the relationships among the parts, the growth of linkages, the process of innovation—formed much of the pattern presented in this book. Their work has given our Center international credibility. More than that, their encouragement and friendship have been the real stimuli and rewards of this enterprise.

THE STRUCTURE FOR
TECHNICAL CHANGE

1 GLOBAL SCOPE

Throughout the world, millions of people engage in activities that make up the technical enterprise, the global effort by which science and technology are advanced. This broad endeavor can be viewed in different ways. It is a unique intellectual activity and a principal expression of the human condition. It is a powerful tool for bringing about economic, social, and political change. And it is a vast enterprise. Most of us rarely think about the people, the money, the organizations at the foundation of these advances. The focus of this book is the vast enterprise that is necessary for creating technical change.

SEEKING THE BENEFITS OF CHANGE

We in the twentieth century seem to be doing something right with regard to science and technology. Those of us fortunate enough to live in the developed, industrialized countries have more choices available in almost every aspect of our lives than did our ancestors. We are enthusiastic about technical advances in transportation, communication, consumer products, and health care, for the most part, though we feel uneasy about pollution and nuclear weapons. We desire and expect continued improvements in our way of life and our

3

economic well-being from the continued success of the technical system.

Not enough is known about this system. Nevertheless, we try hard to obtain benefits quickly, particularly when facing a problem for which technical change may have an answer.

In seeking greater benefits from the technical system through the public sector, we pass legislation, issue regulations, and offer financial incentives. These are all attempts to manipulate the operations of the technical system so that its outputs can match the perceived needs of society. Given how little the general public knows about the system, and even less understands the interactions that occur, these manipulations can be hazardous. We tamper at considerable risk with the goose that lays the golden eggs.

Industry increased funding of its own research and development efforts in the 1970s with an intensity that is a phenomenon of modern civilization. Here, at least, is a purposeful integration of technical activity with economic activity. Cause and effect are better understood within the corporate structure. Interactions with the external technical system in turn exert strong influence on the industrial system. There is evidence that the increasing technical activity throughout the world has stimulated corporate growth. The driver becomes the driven. An element of inevitability, if not predictability, has entered the strategic planning of technology-based companies.

The roles of other important players within the technical system are changing as the system changes. The large research universities conduct far more technical activity than the level of teaching activity or the size of the student body or the number of tenured faculty would necessitate. As academe pursues broader functions within the technical system, it accepts obligations that strain the traditional administrative structure of the university. We are witnessing a move toward a new position for university research.

Government laboratories within agencies such as the Departments of Defense and Energy have relatively clear objectives. Government activities involving support of science and technology intended for general economic benefit are groping toward the most effective ground rules and working relationships. More complex are the impacts of large federal research programs—primarily defense, space, and energy—on private research and economic growth. Direct spin-offs from such government activity to commercial use are infrequent and difficult, and for good reason. Indirect effects, however, may be underrated, and benefits are more likely than negative aspects.

In sum, technical progress is vital to the well-being of our society, which is changing under its own internal pressures and under the well-intentioned actions of public agencies guided by an inadequate understanding of complex phenomena. Even industry, with a direct control of its internal technical developments and their conversion to use, finds its corporate strategies pushed more and more by external technical change.

We can influence the changing technical system to some extent through both public and private actions, but only if we know what the system is and how it operates. There is a good chance that actions of government and industry that are influenced by changes in the technical system can be conducted more effectively with such understanding.

That is, of course, the purpose of this book: to identify and describe the technical system and its interactions with the rest of society, then guide future actions.

The book is divided into three parts, Part I describes the principal players, their current activities in producing technical change, and the pressures that influence developments. Part II shows how the feedback from the growth of the technical system influences the strategies of corporations and governments that made the growth possible. Finally, Part III presents some thoughts that may guide corporate and governmental policymakers.

The process of technical change goes well beyond the conduct of research and development. Some new knowledge, some new insight, or some new combination of known science and technology must be at the core of technical change. But the actions by which we support and conduct R&D, influence the choice of programs, and convert the results to products and processes that are commercially acceptable all make up the *process* of technical change. The term *innovation* connotes a flash of genius. An important point in this book is that technical advance results instead from a conscious strategy that calls for intelligence and a concentration of resources.

Although this book emphasizes the U.S. experience in technology, the internationalization of R&D activity is a key to current pressures on industry and government. Moreover, certain topics of current interest, such as controls on export of technology, are intrinsically international in character. Therefore some discussion on R&D activities abroad is presented.

THE RISE OF INDUSTRIAL RESEARCH

Industrial research is emerging as the driving force behind technical change. Industry's need for technical change is increasing more rapidly than its ability to generate change internally, and this becomes a focus for the strategic plans and growth mechanisms of industry. The internationalization of both R&D and industry are two aspects of this same theme.

The rate of technical advance, our awareness of new technologies in our daily lives, and the intensity of efforts devoted to the generation of new science and technology all seem so different from our views of life in Western society during the fifteenth, eighteenth, and even nineteenth centuries that we have the impression of a discontinuity in the process of technical change. Yet there has been no discontinuity. The nature of technical change has led us steadily and perhaps inevitably from the alchemist's laboratory and the inventions of the Middle Ages to the giant research laboratories of today's industry in a continuous, though accelerating, stream.

Advances in science and technology over the centuries have required progressively more resources—more people, money, facilities, and equipment. Assembling these resources required organization. The inventions that produced the Industrial Revolution made use of existing science and technology. Thus industry of the nineteenth century was structured to apply invention, but not to generate new knowledge. Not until the end of the nineteenth century was a research laboratory funded within a corporation. The rise of industrial research strengthened industry and generated technical change by three mechanisms: organization, feedback, and resources.

1. Industry provided *organization* for the support structure and interdisciplinary contribution necessary for technical advances.
2. Research activity within the user organization accelerated the *feedback* between the use and the generation of new technology.
3. Direct ties between the research organization and the source of economic value—the market—provided the steady stream of financing necessary for *resources* that made continuing technical advances possible.

The growth of industrial research has stimulated scientific and technological achievement throughout the world in universities and government as well as industry itself. As a consequence, every company and every country has become less self-sufficient in generating the science and technology needed for its continued growth. And the need for access to a broader range of science and technology has become pivotal to the strategic business plans of the modern technically based corporation.

MAGNITUDE OF THE TECHNICAL ENTERPRISE

Research and development employed about 750,000 professional scientists and engineers in the United States alone during 1983, accounting for estimated expenditures of $87.7 billion.[1] (This is about 21 percent of the more than 3.5 million professional scientists and engineers employed.)

Almost all of the activity defined as research and development (R&D) occurs in some form of technical organization. This is most often a laboratory, though theoretical work and computers can form a research environment without test tubes and oscilloscopes. Whatever the physical surroundings, there is a support structure for R&D activity that adds from one to three people for every scientist and engineer. These are the technicians, machinists, clerical personnel, and administrative staffs that are just as necessary for the basic research of universities as they are for the product development of corporations. Hence, we can multiply the number of professional scientists and engineers by roughly 2.5 to get an average figure of national or global manpower devoted to R&D. This means that close to 2 million people in the United States are involved directly in the conduct of research and development activities.

This, indeed, is a vast enterprise. It is a considerable economic entity in its own right, a factor well recognized by those concerned with regional economic development. If we assume that the number of people at a single location of R&D activity is on the order of 100, then there are about 20,000 such technical groups in the United States. Again, the precise number is not important, but rather the realization that an enormous network of interactions is necessary to establish effective communication among this large number of locations. Further, the contribution of any single group to the worldwide reservoir of science and technology, and even to advances in its own special area of inquiry, must be evaluated in light of the growing level of technical activity throughout the world.

This complex of activity goes on at universities, in industry, in government laboratories, in private research institutes, and in other specialized institutions. Since this activity incorporates the wide use of grants and contracts, it is therefore important to distinguish clearly between who does the work (the "performers") and who pays for it (the "source").[2]

Table 1-1. Relation between Performers and Sources, 1983.

Funds Provided By: (in millions)	R&D Conducted In:				
	University	Industry	Federal Government	Other	Total
Federal government	$4,900	$20,900	$10,228	$4,300	$40,328
Industry	375	43,700	—	275	44,350
University	1,860	—	—	—	1,860
Other	540	—	—	600	1,140
Total	$7,675	$64,600	$10,228	$5,175	$87,678

Table 1-1 shows who paid whom for R&D in 1983. It reveals industry to play the leading role as both performer and source of funds. Industry conducted about two-thirds of the total national R&D in 1983. Industry funded slightly more than 50 percent of total R&D expenditures. This is roughly two-thirds of the R&D conducted in industry, though the proportion varies radically among different industries. University R&D receives about 64 percent of its support from the federal government and only 5 percent from industry. While the federal government provides about 50 percent of R&D funds, only 12 percent of the effort goes on in federal laboratories.

Table 1-2 shows that scientists and engineers engaged in R&D divide in somewhat the same proportion as the R&D expenditures. There are differences in the cost structures, so that the relationship is not exact, but the dominance of industrial R&D personnel stands out.

It should be noted that the cost per person is not a measure of effectiveness for such broad sectors. Before we applaud the university or berate the government, we should remember that government

Table 1-2. Relation between Funds and Personnel, 1983.

Performer	Funds (millions)	%	R&D Scientists and Engineers (thousands)	%	Cost per Professional Person[a]
University	$ 7,675	8.8	85.8	11.4	$ 89,452
Industry	64,600	73.7	555.9	74.1	116,208
Government	10,228	11.7	66.0	8.8	154,970
Other	5,175	5.9	42.3	5.6	122,340
Total	$87,678	100.0	750.0	100.0	$116,904 (avg)

a. This is the annual cost of operating the technical sector per professional person. It includes supporting personnel, equipment, and services.

Table 1-3. Performance of Basic Research, 1983 (*est.*).

	Expenditures (millions)	Percentage of Total National Effort in Basic Research	Percentage of R&D Effort by Sector
University	$ 5,135	48.4	66.9
Industry	2,050	19.3	3.2
Federal government	1,650	15.6	16.1
Other	1,775	16.7	34.3
Total	$10,610	100.0	

facilities (for example, those of NASA and defense) include highly specialized and expensive equipment, whereas the universities may have inadequate equipment or support personnel. These cost figures are useful only as a rule of thumb to relate funds to professional people. They can be used for comparison purposes only with very similar organizations in the field or, more significantly, to monitor changes in the same organization.

As shown in Table 1-3, universities carry out 48.4 percent of all basic research in the United States, but this is only 66.9 percent of the R&D conducted at universities.[3] Industry, which devotes only 3.2 percent of its total effort to basic research, nevertheless accounts for 19.3 percent of the national total in this category.

GLOBAL DISTRIBUTION OF THE
TECHNICAL ENTERPRISE

In comparing U.S. activity in R&D with worldwide activity, we want to know the following.

- What are the capabilities of our principal trading partners and competitors, the industrialized countries of the Organization for Economic Cooperation and Development (OECD)?

- What is the level and amount of technical activity in the USSR? What is the Soviet Union's ability to generate and to absorb science and technology relative to ours?

- What are the present capabilities and future potential of the People's Republic of China (PRC) both as trading partner and as a provider of technical developments?

- What is the probable total world generation of science and technology relative to that of the United States? (That is, how important is the U.S. technical effort internationally?)

Data available for U.S. activities is far more reliable than that from other countries. However, our purposes here are served adequately by developing reasonable comparisons.

This initial glance at how much technical effort is conducted and in what countries will touch on numbers alone. Clearly, the levels of sophistication and effectiveness are not described by numbers. Still, the number of scientists and engineers carrying out R&D activity in a country gives analysts some measure of the current and potential contribution of a country to the world technical effort.

The OECD includes Western Europe, Canada, the United States, and Japan. The R&D expenditures in countries other than the United States in 1979 were about $65 billion, compared to $57 billion for the United States. Scientists and engineers performing R&D within the OECD came to about 861,000 outside the United States, compared to 621,000 in the United States.[4]

The distribution of effort is, of course, as uneven within the OECD as it is within the 50 states of the United States. The five largest countries—the United States, Japan, the Federal Republic of Germany, France, and England—account for somewhat more than 85 percent of the total R&D activity. Still, we are aware of the world-class technical competence and industrial sophistication of the Netherlands, Sweden, and Switzerland, which, in 1979 had roughly 45,000 R&D personnel and expenditures of $5 billion combined. These figures account for only 3.5 percent of the personnel and about 4.2 percent of the money devoted to R&D, but those countries have become leaders in particular niches of high technology.

The Soviet Union has placed great emphasis on science and technology from its beginnings and even before, since the use of technical advances for economic progress was embedded in the writings of Communist theorists. Data from the NSF *Science Indicators* estimates that there are approximately 1.5 million scientists and engineers performing R&D in the USSR, roughly equal to all the OECD countries combined, including the United States. The Soviet Union leads all other industrialized nations in such indexes as the percentage of gross national product devoted to R&D and the number of scientists and engineers per 10,000 people employed. Despite this, there is no great concern about competition in fields of high-technology products from the USSR, nor is the Soviet Union considered to be the leader in any substantial number of technical fields. Americans' current concerns about "international competitiveness" are obviously more focused on Japan, with half the number of R&D professional personnel, than on the USSR with twice as many. This is an important perspective to keep in mind when examining relative data on technical effort in different countries.

Measures of R&D inputs are only guides to future performance. They tell us nothing about effectiveness, about the structure of the system in which R&D is conducted. The Soviet system is simply not designed for the effective generation and use of technical change in the civilian sector. This is not necessarily true for military use, which follows different criteria for the planning, conduct, and integration of R&D. There is not the slightest evidence that the situation is likely to change in the foreseeable future.[5]

The People's Republic of China is another story altogether, and could well be a vital and powerful factor in technical change 10 years from now. The Communist system there has much of the restrictiveness seen in the USSR, but there are many signs of a willingness to be more flexible and far more pragmatic. The Chinese may not be able to institute economic incentives along the lines of venture capital or tax advantages, but their innovativeness may produce substitutes. They have established a prototype School of Business Administration at Dalian[6] that will at least provide an understanding of the features needed to adopt and manage change. There is every reason to expect that structural changes to improve the overall productivity of R&D— both its conduct and transfer to use—which can be instituted without requiring major political changes, will in fact move ahead.

China is in the early stages of modern science and technology, and apparently has on the order of 500,000 scientists and engineers in R&D, about 75 percent the level of the United States.[7] China's rate of growth is faster, its intellectual capacity high, and its intensity obvious. We must expect that in 10 years it will have more R&D manpower than the United States within a technical structure that is reasonably effective by any measure. The implications for China's economic growth and competitiveness in international markets are worth serious consideration.

It is difficult to get reliable data on total R&D activity throughout Africa, Latin America, and Asia outside of Japan. There is a considerable base of technically trained personnel in Egypt, India, many of the "newly industrialized countries" (NICs) such as Korea, Taiwan, and Singapore, as well as in Mexico, Argentina, Brazil, and Chile. The number actually engaged in R&D activity is much less than the number of people trained in science and technology.

Nevertheless, estimates are available that can give us a measure of R&D activity in the Third World. The total number of scientists and engineers conducting R&D in 1980 was approximately:[8]

Latin America	80,000
Africa	
non-Arab countries	30,000
Arab countries	45,000
Asia—excluding Japan	400,000
Total	555,000

Combining all the sources of data quoted, and assuming the same growth rates from 1980 to 1984 for the OECD and Third World countries as in the several years preceding, we can assemble the picture of worldwide R&D activity shown in Table 1–4.

Thus, the United States accounts for about 18 percent, or roughly one-sixth, of professional R&D personnel in the world. The industrialized countries of the OECD, including the United States, constitute about 40 percent of the world effort. Since the OECD activity takes place as part of a generally more effective industrial research system, has developed feedback between user and generator, and

Table 1-4. Worldwide R&D Activity, 1984 (*estimated*).

United States	750,000
OECD (excluding U.S.)	960,000
USSR	1,500,000
China (PRC)	500,000
Third World	560,000
	4,270,000

maintains a critical technical mass in most areas of economic growth, the impact on worldwide technical change from the OECD activity is undoubtedly greater than the 40 percent share based on personnel alone.

Those numbers hint at the magnitude of the worldwide technical enterprise. Allowing the crude figure of $100,000 as the cost per professional person annually (about 15 percent less than the amount in the United States) means over $400 billion of technical activity is conducted worldwide each year. There is probably less investment in advanced equipment and buildings overseas, but this may be made up in part by more support personnel per researcher. Whatever the exact figure, it is indeed a vast enterprise.

It seems obvious that any activity of this size has impact and has an organized structure. The effectiveness of the overall system and of any single technical organization within the system is affected by the nature of the internal structure and by the relations between the technical process and other activities within society. These topics make up much of the substance for this book.

NOTES TO CHAPTER 1

1. National Science Foundation, "Science Indicators, 1982," NSF 83-1, 1983.
2. All data referred to in the remainder of this section, unless otherwise indicated, derives from National Science Foundation, "National Patterns of Science and Technology Resources—1984," NSF 84-311, 1984.
3. The National Science Foundation's distinctions between basic research and R&D, used here, are stated as follows in "Academic Science Engineers: R&D Funds Fiscal Year 1982," NSF 84-308, 1984, p. 143.

   Basic research is directed toward an increase of knowledge; it is research where the primary aim of the **investigator** is a fuller knowledge or understanding of the subject under study rather than a specific application thereof.

Separately budgeted research and development (R&D) includes all funds expended for activities specifically organized to produce research outcomes and commissioned by an agency either external to the institution or separately budgeted by an organizational unit within the institution. *Include* research equipment purchased under research project awards from "current fund" accounts. Also, *include* research funds subcontracted to outside organizations. *Exclude* training grants, public service grants, demonstration projects, and departmental research expenditures that are not separately budgeted. Also, *exclude* any R&D expenditures in the fields of education, law, humanities, music, the arts, physical education, library science, and all other nonscience fields.

4.   *OECD Science and Technology Indicators* (Paris: OECD, 1984).
5.   J.R. Thomas and U.M. Kruse-Vaucienne (eds.), *Soviet Science and Technology* (Washington, D.C.: George Washington University, 1977).
6.   This was in 1979 at the initiation of Jordan Baruch, then Assistant Secretary of Commerce, under a U.S. team headed by William Dill, then Dean of the Graduate School of Business Administration at New York University, now President of Babson College.
7.   Leo Orleans, *Training and Utilization of Science and Engineering Manpower in the People's Republic of China*, Background Study No. 5, Library of Congress, Washington, D.C., October 1983 (Government Printing Office No. 21-8570).
8.   *Statistical Yearbook* (New York: UNESCO, 1983).

2 UNIVERSITY RESEARCH
Beyond Education

There is a mystique about university research, much of it understandable. There is also a misconception about university research that does the university a disservice and can confuse or inhibit public policy. It is necessary to clarify thinking about this segment of the technical enterprise for reasons beyond the simple statement of facts. To many people, research *means* university research. This is clearly not correct in terms of numbers of dollars and people employed, and it is not correct in terms of function or objectives.

More important, many in government and in the academic community appear to consider national science policy as good if it increases funds for university research, bad if it does not. Apart from the foolishness to which any such one-issue reasoning leads, this model of the process of technical change contradicts current experience.

The misconceptions are thoroughly understandable. The university was almost the sole source of scientific research and of new advances in technology prior to the initiation of government laboratories in the nineteenth century and the growth of industrial research in the twentieth. The university remains the primary source of basic research and hence of Nobel prizes. In addition to this measure of success, the sudden realization of commercial possibilities in genetics and biotechnology has focused public attention on the great advances achieved by university researchers in these areas.

Nevertheless, we must be aware of three points. First, university research is a small part of the total technical enterprise: Technical activities at universities constitute under 10 percent of total R&D in the United States, and university basic research is less than 50 percent of the total U.S. effort in basic research. Second, the bulk of university R&D is conducted by the relatively small number of research universities that have taken on the obligations of research management as a new university function. Third, university basic research, while quite possibly the most important long-term activity in the national R&D spectrum, is not the driving force for the process of technical change that it may have been prior to World War II. That role is now served by industrial research.

THE U.S. UNIVERSITY RESEARCH SYSTEM

We use the term *university* loosely. The National Center for Education Statistics (NCES) defines a university as an institution granting doctoral degrees which has at least two first professional programs (medicine, dentistry, law). According to unpublished data from the NCES, the number of postsecondary institutions in the United States in 1983–84 can be arrayed as follows:

Two-year colleges	1,271
Four-year and over colleges	1,898
Universities	115
Total	3,284

This is the base of higher education in the United States. Apparently 2,200 of these institutions employ scientists or engineers, but the NCES statistics do not distinguish between teaching and research appointments. Research activity is concentrated in the graduate schools, principally in those institutions granting the Ph.D. degree.[1]

According to the National Science Foundation, the 50 largest research institutions accounted for 60.9 percent of all academic R&D and 63.4 percent of all federal funds for this activity in 1982. The first 100 institutions had average R&D expenditures of $60 million each, while the second 100 averaged $10 million. If we set aside Johns Hopkins with its $199 million from the federally funded Applied Physics Laboratory, then the largest R&D effort is that of the Massachusetts Institute of Technology, with a total of $192 million, of which $157 million comes from the federal government. By contrast, the two-hundredth institution was the Worcester Polytechnic Institute with total R&D activity of $4.01 million, of which $1.61 million was federally funded.

For all practical purposes, R&D stops with the top 500 institutions, leaving roughly 2,800 schools and colleges as essentially teaching institutions. The institutions making up the last 300 of those conducting R&D comprise 3.4 percent of total academic research, or $250 million, of which $140 million is federal funds. MIT alone receives more federal R&D support than do these 300 institutions combined. These are not at all insignificant or unheard-of schools. They represent a wide spectrum of our very best and very well-known colleges and universities, including Southern Methodist University, Clark, Wellesley, William and Mary, Smith, Claremont, Amherst, Bryn Mawr, Reed, Haverford. Even the last 100 include Virginia Military Institute, Grinnell, Goucher, DePauw, Carleton, Franklin and Marshall, and Oberlin—all great schools in the best academic tradition. They are simply not involved in major research programs, but they do combine some research with the principal teaching activity.

Thus research institutions are practically synonymous with doctorate-granting institutions, which accounted in 1982 for 98.2 percent of all academic research and 98.3 percent of federal funding of academic research.

The geographic distribution is of economic and political interest, but with few surprises. The rim formed by the West Coast, North Central States, and New England and the East Coast make up a significant portion of academic research. Practical distribution of funds is by state, not region, however. The state figures deserve comment because of their significance for public policy. If we consider the success of the states with the largest academic R&D funding in terms of how many federal dollars are granted for every dollar of nonfederal funds in academic research, we have

Massachusetts	$3.61
California	2.52
Pennsylvania	2.29
New York	2.18
Texas	1.26
Average	$2.37

The extent to which the ratio of federal to nonfederal funding represents political influence, technical excellence, or choice is too much of a diversion from the thrust of this book. The states that do the *least* R&D at academic institutions are, not surprisingly, the least populous. They rank high in academic R&D per capita. But the lower absolute amount of R&D attracts less federal support per dollar of R&D invested than in the highest ranking states ($1.44 versus $2.37). This reflects the level of effort needed to contribute significantly to technical progress and thus attract federal funds.

THE UNIVERSITY AS A MANAGER OF RESEARCH

The very phrase *manager of research* offends the traditionalist, who pictures academic research as Einstein at his cluttered desk. The traditionalist must consider that any university administrator who did anything other than provide total support for Einstein would be exercising extremely poor management, and that Einstein did not take on responsibility for operating a linear accelerator, coordinating the inputs of 10 junior faculty or nonfaculty researchers, purchasing computers, assigning computer time, and so on.

The "management" aspect of academic research arises from concentration of research activity, and this is most often related to federal funding. This is seen more vividly if we return to the comparison of MIT with those institutions that ranked from 201 to 500 in total research during 1982.[2]

Thus, the concentration of $35 million of nonfederal funds at MIT attracts $4.49 of federal money for every dollar of nonfederal funds. The larger amount of diffused research activity throughout the 300 institutions attracts a little more than matching federal grants, namely $1.29 for every nonfederal dollar.

This data serves only to highlight the principal elements that create the role of the university as a research manager. These are

- Some minimum size of effort (the institution ranked 200 spent $4 million on R&D in 1982, the institution ranked 500 spent only $18,000)
- A concentration of resources
- Attraction of federal funds

NSF's statistics demonstrate that, with regard to research, there is a three-tier pattern among postsecondary institutions in the United States. The first tier is composed of schools and colleges that are primarily teaching institutions and do little or no research. The second tier is made up of the substantial number of colleges, roughly 300, that follow the traditional picture of academic basic research in which some research is conducted by faculty members, and research support is solicited for the work of individuals or a small group of researchers. The top tier is an assemblage of perhaps 200 institutions which are very much research oriented, which solicit the large grants that require for their performance the employment of scientists and engineers who are not faculty members, and where the level of research activity may have no simple correlation with teaching, faculty size, or number of students.

In practice, then, the great portion of expenditures on university research is accounted for by institutions which have structures for soliciting and monitoring funds for this activity, and almost always embody some organized structure for the conduct of major programs. Since there are substantial numbers of people engaged in research at these institutions who are not faculty members, this adds a different dimension to the nature of university research—which is a challenge to effective university administration.

The 200 research institutions have become managers of academic research. This is an intriguing feature, and certainly one that is not emphasized in general discussions of the "natural" evolution of research as a close partner of teaching. In 1982 the 100 largest research institutions accounted for 83 percent of all academic research and 84 percent of federal funding of this research. These universities are now engaged in managing over $6 billion of research. They have solicited and have accepted the responsibility for the conduct of a broad technical enterprise. Stating this somewhat differently, society has delegated to this group of academic institutions a number of missions, including but not limited to undirected basic research, which require the allocation and management of resources and the conduct of very substantial organized technical activity.

The management and organization which are in the very nature of large programs may seem contrary to an idealized view of academic research. Because of the number of large research programs and the total level of research activity carried out by the major academic research institutions, however, organization and management are

required. Such programs are very much part of the educational process in the broadest sense though related only loosely to formal
teaching. Research management on this scale represents an aspect of
universities that has crystallized in the second half of this century
and deserves a close look.

THE NATURE AND FUNCTIONS
OF UNIVERSITY RESEARCH

The university function of "research manager" is not incompatible
with the essential university values and customs, but it does not mesh
easily with them. That is an important source of strain within the
modern university.

"Research management" conveys the image of purpose, hence of
applied research or development. "University research" is most often
taken to be synonymous with basic or fundamental research, hence
with general knowledge, unrelated to an end item or plan. Obviously,
both implications may hold in general but can be incorrect in practice. One can "manage" basic research by allocating resources and
support that permit it to proceed most effectively. Basic research can
be purposeful and can be of immediate use in guiding technical progress toward a specific product or mission.

To understand both the images and the apparent contradictions,
we must look at the core of university traditions. The freedom to
think and to express one's thinking in teaching what we know leads
naturally to asking questions about what we do not know. It is therefore not surprising that university education is closely accompanied
by university research, and particularly by the pursuit of basic
research.

This freedom of inquiry in combination with the tenure system is
what makes research in American universities unique. It is the most
effective system we know for the encouragement and support of
the individual researcher. Freedom of inquiry leads directly to the
strength of individual research. The tenure system assures the researcher economic security and hence the freedom to pursue challenging areas of research without fear of penalties. Tenure is granted
by one's peers, normally represented by one's colleagues in an academic department. Thus a physicist is granted tenure by physicists,

who can best judge the person's work, a mechanical engineer by mechanical engineers, and so on.

So freedom of inquiry, tenure, and departmental structure are related by the traditional university foundation. The modern research university superimposes on this foundation large-scale research programs and mission-oriented centers that call for contributions from researchers in different departments, usually working toward a coherent objective, and coordinated by some form of research management. It is organized interdisciplinary research. All of the incentives and all of the traditions of the university structure would appear to oppose these new research mechanisms. Many academics and university administrators worry that any major university enterprise which is conducted outside of departmental controls can modify the tenure system, which might in turn threaten the truly fundamental freedom of inquiry.

If freedom of inquiry and basic research were rigid concepts, there would be unbearable strains in the university system. They are in fact relative concepts, and this provides the flexibility necessary to accommodate changes in university functions and obligations without endangering the basic university strengths.

Freedom of inquiry is equivalent to freedom of choice. An immediate practical restriction is that there are finite resources available. The allocation of these resources cannot possibly be decided by the free choice of individual researchers. Some researchers are simply more competent than others, and some technical areas are growing more rapidly than others. Some research areas call for large resources, others for very little. A broader authority, whether that of consensus or of formal administration, limits the freedom of inquiry on highly pragmatic grounds by decisions on the allocation of funds, space, assistants, and so on. Teaching itself is a restraint. We may joke about letting teaching interfere with research at universities, but it *is* a limitation on time and on selection of new personnel.

Other objectives of society compete with freedom of inquiry in the performance of the university, which is itself an instrument of society. One is a national crisis, the clearest example being war. The response of the U.S. university system in World War II was perhaps one of its finest hours, and certainly a dramatic statement on the position of the university system as a vital participant in society. Less dramatic, but of longer duration, are the current concerns with economic growth, which bring increasing attention to potential university contributions.

The correct university response to these varied pressures is that it maintains the freedom to identify programs of value and interest and to control directions of research, taking into account whatever limitations exist. This is fair enough. The real freedom of choice, then, lies in deciding which conditions are acceptable. The many real-life restrictions and conditions, important or trivial, under which university research is conducted have been accepted freely by faculty and administration.

Basic Research: Directed versus Undirected

Basic research provides understanding of some natural phenomenon, some cause-and-effect relationships, some property of materials. When this activity is conducted because of the intellectual curiosity of the researcher, it may be called *undirected basic research*; when it occurs because the activity will contribute to a broader mission-oriented objective, it is called *directed basic research.*

There is an old joke in the research laboratory that goes, "If I think of it, it's basic research; if you think of it, it's applied research." This is psychological and it is real, but it has nothing to do with the nature of the work or the results.

The large research programs at universities and the operations of research centers are financed primarily by government agencies, and increasingly by industry. These sponsors often have broad missions to accomplish, some purpose for funding university research. Nevertheless, the bulk of university activity which they sponsor is basic research. It is *directed* basic research as far as the sponsor is concerned.

A researcher with a concept for basic research may seek a sponsor with a related mission. If it fits a university research center, so much the better. If it is in biology, the researcher might seek funds from Merck or Hoffman-LaRoche. If it is in materials, funds might be solicited from Alcoa or General Electric. To the researcher, it is undirected basic research. To the sponsor it is directed basic research.

Suppose a major sponsor indicates that money is available for basic research in particular areas of science or engineering, often for very specific programs. Predictably, research interests arise to meet the funds. But this in turn stimulates research interests.

The directed basic research of one generation becomes the undirected basic research of the next. At any moment, the spectrum of basic research at universities is a compromise between the priorities of funding sources and the interests of the researchers. The diversity of funding sources combined with the diversity of universities has been able to accommodate both the scientific needs of the funding sources and the independence of universities to explore new directions. Basic research and freedom of inquiry are alive and well, and have demonstrated their compatibility with research management as a new function, at the major research universities.

One example where the university system has worked well with research centers and interdisciplinary research is the experience with the Materials Research Centers (MRCs). These were established by the Advanced Research Projects Agency (ARPA) of the Defense Department from 1959 to 1961. Eventually, 15 MRCs were established at institutions that included the universities of Chicago, Illinois, Maryland, as well as MIT and Cornell. They were funded at approximately $1 million to $5 million annually.[3]

The MRCs were assigned the goal of increasing basic scientific understanding of the structure of materials. Advances in nuclear reactors, jet engines, supersonic aircraft, and missiles required materials that would be strong at high temperatures, resist nuclear radiation, and operate in corrosive atmospheres. Empirical advances were slow and limited so that a major scientific thrust was necessary to provide wholly new possibilities.

The Centers were interdisciplinary in the sense that researchers from different disciplines made contributions and were exposed to other disciplines. They advanced knowledge of materials, stimulated research, and produced a generation of researchers with a strong foundation in advanced materials science. Each researcher kept the affiliation with the home department. The diverse research interests were sheltered under the MRC umbrella. Basic knowledge was gained that aided the major national programs and helped to develop graphite fibers, cermets (ceramic-metal combinations), high-strength alloys, and other new materials.

The MRCs did not require truly interdisciplinary research, where a project in one field is necessary to complete a project in another.

Interdisciplinary research calls for coordination of efforts, for advance planning, for feedback—activities not familiar to the university researcher and not facilitated by the university structure, but typical of the newer research centers at universities.

The MRCs were successful because they built upon university strengths that already existed. The new function of research management must combine coordination of research with the independence of action traditional to the individual university researcher.

IMPORTANCE OF UNIVERSITY RESEARCH TO THE UNIVERSITY SYSTEM

The research budget figures suggest that the top research universities, which have become managers of research, differ somehow from the idealized view of the university. There is no way that the university can make its optimal contribution to the technical enterprise unless the university community accepts fully the philosophical implications of superimposing the management of research upon traditional university values. The rest of society, especially government and industry, must understand the nature of changes in university functions and processes that have evolved in order to develop the most effective working relationships with the academic research system.

We should first ask, Are there any serious problems? There probably is not, with regard to the continuing research output of the American university system, its leadership in most current areas of basic science, and its ability to stimulate new business ventures through people and ideas. In answer to those who "view with alarm" any of several current trends that might affect the university system, university research in the United States is strong and performing its functions in the technical enterprise with considerable effectiveness.

Beyond the Applause

The successes of the university system deserve praise, but there are pressures on the university system that must be considered to avoid future difficulties.

- External expectations do not always take account of the differences among universities, the nature of university research, and the interests of university researchers.

- Expectations that measure the obligations of research management in terms of traditional university values can limit its effectiveness and create strains within the university.

- Issues requiring public action raise questions about university procedures:

  Are there inadequacies in university research equipment?
  Is the university able to attract adequate faculty?
  How can regional economic development build upon university relationships?
  How will controls on exporting technology affect university research?

The single thread joining these considerations is the need to understand and accept the nature of the university research system as it exists. Too often arguments are advanced both inside and outside the university as if the research activity is a simple adjunct of teaching. This was true in the past and is largely true today but does not adequately address the growth of research management. There is a tendency to consider current procedures in terms of older and idealized values, a tendency that is sometimes nostalgia, most often unfamiliarity, and occasionally hypocrisy.

This discussion is not new, though the particular forms of university research are different today. Perhaps the clearest exposition of the issues was made by James A. Perkins, then president of Cornell University, in the Stafford Little Lectures at Princeton in November 1965.[4] He emphasized the responsibility of the university for advancing knowledge in its three aspects—acquisition, transmittal, and application. These translate into research, teaching, and public service. Dr. Perkins was examining the growth of "big science" at universities pushed by federal government funds, a situation dramatized by Clark Kerr who, as Chancellor of the University of California, referred in 1963 to the growth of the "federal grant university."

Perkins considered how much and what kind of research and public service would maintain the integrity, the strength and coherence of universities. He favored action in any one area of the university's three broad functions that strengthens the other two. The research function is, in the ideal case, embodied in the teacher–researcher. When the professor teaches and conducts research, there is rarely a problem. Where there is more of one activity than another, there should at least be coordination of the two, so that the teaching relates to the research within the department for example.

With regard to the university's role in public service, the "application of knowledge to the problems of modern society," Perkins had a classically pragmatic criterion: The university's competence is in knowledge, not operations. Involvement in public service should be limited to "advice on how to do something" not "assistance in doing it."

Most people probably would agree with these criteria and believe that the university system is in line with them today, 20 years after Perkins's lectures. Taken literally and with reasonable common sense, however, these criteria do not apply to the major research management obligations of the top 50 to 100 research universities. The cautions about public service would limit the consideration that some universities are giving to their role in start-up ventures and mechanisms for exploiting their research.

The university is a marvelously flexible and resourceful institution. Today research management is contributing to the university's strength, and commercial exploitation of university research can be kept within the scope of public service, albeit with some buffer to keep research faculty on the campus.

To support and exploit the university's expanded role we need to acknowledge that the university system is carrying out new functions, not simply expanding those of the classical university, and the great diversity within colleges and universities must be considered in public policy.

ESTABLISHING PRIORITIES

Consider the question of research facilities and equipment. It has become an article of faith for university administrators, legislators, and even industrial leaders to state that university research facilities are inadequate. Various studies indicate that this is probably true, but what does this mean?[5] There are, consistent with the three-tier system of colleges and universities, three broad categories of scientific equipment: (1) equipment needed for teaching, (2) equipment needed for research, and (3) specialized facilities for advancing frontiers in particular areas.

Most institutions of higher learning are concerned only with the first category. General upgrading of research facilities is of concern to the 500 institutions engaged in research, but the degree of upgrading called for is surely different for the last 300 than for the first 200. Advances at the frontiers of technology call for capital investments in facilities that are available only to the major corporations in the field and a few university groups. Almost by definition, most of the large research universities which conduct research in a particular field have inadequate facilities.

Now what appeared at first to be a simple question of the adequacy of university research facilities turns out to be subjective, fragmented, and a matter of national policy. Suppose 10 major universities become completely up to date with the most advanced facilities for microelectronics research while another 190 research universities do not. Is the American university research system overall inadequately equipped for such research?

The relation between research and teaching leads to the same questions of priority and policy. What level of research equipment is needed to provide desired graduate education in each of the separate fields of mathematics, physical and biological sciences, and engineering? Should each of the 500 research institutions have equal research facilities in each of the separate disciplines? Is this more important than upgrading the equipment needed for teaching in the predominantly undergraduate institutions that do little or no research?

It seems obvious that research equipment and facilities at U.S. colleges and universities can be upgraded. It is not at all clear what priorities should be established among the different needs and functions of university teaching and research. It is even less clear what the role of government should be and, assuming there is a desire to use some government funds, what mechanisms and criteria should be established for their allocation.

The anatomy of just this one issue shows why an understanding of the university research structure is necessary to an informed discussion. It brings out a further point that is present in almost every consideration of the university research system: the difficulty, and very likely the undesirability, of separating the two functions of research and education, Perkins' "acquisition and transmittal of knowledge."

Much of the concern, both philosophical and practical, with keeping the university strong while expanding its contributions lies with the coexistence of teaching and research, and its consequent public service. What Perkins did not stress, probably because it seems self-evident, is that teaching is *primus inter pares*, first among equals. Teaching is the principal, almost unique, function of the university and its historical reason for being. Other institutions can and do conduct the other functions. University research takes place in a system which is structured for, and must give the highest priority to, teaching.

This is a source of strength for the traditional case of the faculty member who combines teaching and research. It is a source of limitation for the management of research on a larger scale. There are solutions, but there are strains.

True to its historical evolution, university research is organized primarily along the lines of university education, by departments such as chemistry, physics, or mechanical engineering. The appointments of faculty, the allocation of resources, the availability of graduate research assistants come within the operating procedures of individual departments. This does indeed, almost by definition, tie the research efforts to the educational program of each department. Experience shows that this structure, incorporating tenure granted to faculty members by departments, has been enormously successful in stimulating fundamental research by the individual researcher in the finest academic tradition—independent, long-range, trail-blazing scientific activity. It has been an effective coupling of research and education.

As we have seen, university research expanded far beyond what would be the level of activity expected from a faculty whose size was determined principally by the size of the student body. The funding came necessarily from sources not responsible primarily for education. The objectives of the research and the missions of the university broadened inexorably to encompass a range of constructive and vital objectives of society other than that of education.

Whatever else we may conclude about these developments, one statement seems evident: If we set forth an organizational structure that would most effectively initiate and conduct a large-scale research and development activity, serving multiple interests, involving responsibilities for linkages and transfer, we would almost certainly develop a structure quite different from one that is among the disciplines and under the authority of university departments. The optimal organization of university research that serves many missions and objectives, even though the activity is very largely basic research, is not the optimal organization of university research that exists primarily to strengthen education.

This is not necessarily a problem, or at least not an immediately critical problem, but it does cause strains within the system. However, neither the university nor society is completely free to resolve these strains separately. If some combination of federal government, local political units, and industry were to propose major modifications in university research structures and financing to serve better a range of objectives, including economic development, defense research, and public health, we might diminish the unique strength of a graduate education system tied into independent research. By the same token, if the university system were to scale down the level of research activity and restrict their objectives nostalgically, to the pattern that existed prior to World War II, in order to preserve a system wherein education and research are most compatible, society could not permit it. University research today has become far too important to the overall technical enterprise, and consequently to the economy and national security, to permit any substantial or discontinuous pull-back.

There is a certain odd stability in the strains of the existing structure of university research resulting from the obligations of research and research management within an educational institution. Society accepts inefficiencies in the university research structure in order to obtain the inputs and attention of university researchers. Universities accept linkages and involvement with society, with its added conditions, in order to obtain greater resources and play a constructive role in current issues.

ASSURING THE EFFECTIVENESS OF UNIVERSITY RESEARCH

The ability of the university to generate new science and technology and to integrate these activities with the broader technical enterprise increasingly focuses on the research centers, institutes, and programs which account for the principal growth of university research and, thus, of university research management. There is insufficient data on these centers of all forms to provide accurate quantitative summaries. A short review by John Walsh gives some clues as to their number and significance.[6] The total R&D conducted at MIT during 1984 was separated into $81 million within departments and $137 million within centers. Of the 7,500 nonprofit research units listed in the 1984–85 *Research Center Directory*, most were related to universities.

To generalize the situation in very rough terms, the many centers

- Constitute most of the R&D activity in the large research universities
- Are interdisciplinary in subject matter and intent

Both characteristics are intrinsic reasons for establishing such centers. When the research objectives call for activity from different disciplines, a "neutral" center serves to emphasize the objectives and to stimulate cooperation from the separate researchers, assets which might be more difficult to develop if the program would be located within a single department. Second, when substantial amounts of money are involved, the use of a separate organizational structure is a clear administrative action that can separate year-to-year commitments for the salaries of research personnel associated primarily with the center from the permanent salary commitments for research faculty with appointments primarily within established departments. Thus, a center may simply form a convenient mechanism by which funds can be earmarked for a particular research activity and personnel can be appointed without formal relations to traditional department activity.

Depending on the degree of interdisciplinary efforts required, a center can be located within a department. Depending on the conditions of funding, a center can engage in either directed or undirected basic research. The growth of large research programs at the major research universities has occurred principally in interdisciplinary centers emphasizing directed basic research.

The increase in university research through such centers has particular significance for our overall concern with the technical enterprise generally. Broadly speaking, the mechanism of research centers is a workable structure that does act to bring together different contributions related to a mission or interdisciplinary objective, that serves as a useful accounting tool and a convenient unit for assigning administrative responsibility, and provides a basis for appointment of research personnel distinct from department appointments of teaching faculty.

The status of research personnel and the nature of interdisciplinary research raise questions about the effectiveness of research management at universities. This is very often a question of what is expected. The Materials Research Centers worked well because their objectives called for a broad increase in basic knowledge of materials science, plus the production of trained and interested researchers. These objectives match the university's strengths.

Do we want more? Occasionally yes, and more frequently as we encourage the growth of research centers intended to strengthen industrial research. The National Science Foundation has embarked upon a major program to establish Engineering Research Centers[7] that will involve a "team effort of individuals . . . possessing different engineering or scientific skills," where "the focus of the Center should be on a major technological concern of both industrial and national importance." Ten Centers were started in 1985. This program is expected to reach a level of $100 million or more annually, with the financial support planned for shifting to industry after five years of operation for each Center. Other centers established within the past decade stress similar aspects of group research, coherent contributions to defined research goals, and linkages with specific needs of government or industry. All of them include the implicit commitment to continuity. These include the Catalysis Center at University of Delaware, the Robotics Institute at Carnegie–Mellon University, Laboratory for Laser Energetics at Rochester University, the Ceramics Center at Rutgers University, the Center for Integrated Systems at Stanford University, and many others.

The growing resource of large-scale university research is valuable because of its impact on the main university assets—students and faculty. It draws graduate students into the research, thereby providing financial support and strengthening the students' research experience and their interests in current fields. It stimulates faculty interests and provides material for teaching advanced concepts and techniques.

Moreover, the magnitude of university research effort devoted to these larger programs raises the general level of basic science and engineering, and raises other expectations. We really do want to strengthen and advance the technical base of industry and the mission agencies of government. We really do want to promote regional economic development. There is a very real pressure to improve the nature and delivery of health care. And there is a steady pressure to strengthen international competitiveness.

All of these values can be added to by large-scale university research programs. That places a very serious obligation on the universities to provide research management that devotes attention to producing those values and not simply to accept whatever results emerge after due arrangements for absorbing graduate students and supporting faculty research.

Tenuring Research Positions

The critical factor is research staff. The research activity within a mission-oriented center or large university research program is conducted by a mix of faculty and staff with research appointments. Such research personnel are often postdoctorate professionals who accept these positions to continue the research interests developed during their doctoral program, or who have entered into a temporary "holding pattern" prior to identifying a more permanent career path, most probably at a university. There are also research personnel who are concerned with the subject matter of the research program, no longer considered to be postdoctorates, but simply members of the center or laboratory without any faculty appointment or tenure.

Concern for the mission of a center or major program should include attention to two characteristics of the research staff: opportunities for stability and growth, and compatibility with the interdisciplinary nature of the subject matter. Neither of these is best served when the only permanent appointments available are restricted to regular faculty, and when those appointments are made only within departments identified by the separate disciplines, physics or mathematics, say.

It is in these regards that the university and society have compromised. All interests are served by strengthening the training of competent graduates and encouraging the research interests of faculty in cooperative or interdisciplinary areas. Thus, there is genuine reluctance to depart from the well-tested structure of departmental control of permanent appointments. But the number of those permanent appointments is based on the teaching needs of the departments, the university preferring to rely on "hard money" from tuition to underwrite tenure, rather than "soft money" from research contracts that may not be renewed. Even long-term support for, say, three to five years, serves to provide greater effectiveness in research planning than to encourage permanent appointments.

Further, the strong departmental role exercised by those with regular faculty appointments favors a form of cooperative interdisciplinary research rather than a coordinated activity. This is satisfactory for raising the general level of basic knowledge in an area, but it may not produce optimal results in a mission-oriented center or program. This will be a more serious concern in the new engineering centers (robotics, for example) than in the earlier science-oriented centers (catalysis at the University of Delaware) or the Materials Science Centers. It will be an increasingly important factor in developing constructive linkages with any external objective, such as stimulating regional economic development.

To fulfill its new role as research manager, the university must strengthen that management. Tenured research appointments tailored to the mission of the center or program should be offered. This would lead to a core research group whose careers and objectives are tied to the research mission. Tenure can only be considered in cases where the mission is sufficiently broad and deep that it will not fade away in three or four years, which is precisely the basis for discipline-oriented departments. Nevertheless, the substantial funds being committed to university research and the increasing integration of that research into objectives of society other than education and basic knowledge are sufficient reasons for universities to reexamine old shibboleths and ways of thinking.

There are other potential advantages to be derived from offering tenured research appointments. Much concern has been expressed in recent years about the difficulty of attracting competent young faculty against the attractions and high salaries of industry. There is good reason to believe that this is not only a question of salary alone, but of status. A tenured research appointment can be a partial answer, particularly for those who want work in an academic atmosphere but do not have the desire to teach. Stability may not completely overcome salary differences, but it would surely help.

Can the university system absorb tenured research appointments? An oblique answer is that many educational institutions have fallen on hard times and that tuition no longer is the "hard money" sufficient to support tenured teaching faculty. A more direct answer is that tenured research appointments can be made if they relate easily to the capabilities and tradition of the institution for research. Research today is a business with an income, as is education itself. When a university has $20 to $50 million of research income year after year, it is not too risky to commit $1 to $2 million in tenured research positions. The trick, of course, is to select the appropriate positions. There really is no substitute for good research management to match people, areas of inquiry, and sources of funds.

The extreme case would be a wholly research university with all research appointments. The closest example of this is Rockefeller University, an institution so unusual that it is not really an example at all. Yet there are surely lessons to be learned from an institution with 300 faculty and 100 graduate students which has produced 19 Nobel laureates. In 1982 it conducted $46 million of R&D, of which $21 million was government funds.

The Rockefeller Institute for Medical Research was a pure research activity from its start in 1901 until 1953.[8] It then converted to a graduate institute and, in 1964, became the Rockefeller University. It offered degrees in several areas related to its competence and attracted a student body. The circumstances were unusual, since the fields were few, the faculty outnumbered the students, and the students received a stipend for their efforts. Not surprisingly, a highly select student body has resulted. The institute became a university partly to formalize the educational intent of its charter and to carry through an obligation to transmit knowledge for future generations, and partly to prevent isolation and encourage the stimulation of ideas that come from explaining concepts to someone else and answering the questions that arise in the process of teaching. Teaching and research are indeed a two-way street, with research strengthened by teaching as much as the other way around.

A crucial characteristic of Rockefeller University is that it is truly interdisciplinary. It has laboratories, not departments. Thus, we have at least one example of an institution based first on research, but incorporating a devotion to education in the broadest sense. Its appointments are based on research, and the curriculum is based upon the teachings of individuals drawn from various research interests, not departments in the traditional sense.

Clearly the operations of a university with a focus on research require a different view of funds than the traditional view based primarily on teaching. Rockefeller University has always had the wealth of the Rockefeller family as a base. Still, while the early years were wholly financed from the family endowment, that endowment today finances less than half of the current operating costs. Increasingly, external sources, private and public, provide research support. The university's existence and growth depend almost exclusively on its excellence and effectiveness in its selected research areas. It is surely one of the brightest lights in our technical enterprise, though with a structure and emphasis that is unique among universities.

The major research universities, by contrast, are producing hybrid structures for research. The principal growth in university research is in mission-oriented centers and programs, but permanent appointments and faculty control remains with departments. This derives from the primary emphasis on teaching, even though the income from tuition provides a steadily decreasing share of the university's income. Typically, tuitions yield approximately 20 percent of operating costs for the top research universities. (At Massachusetts Institute of Technology tuitions pay for only 14 percent of operating costs, at New York University 30 percent.)

Thus, both the increased role of university research and the decreased share of finances from tuitions place greater pressures on the management of university research to optimize their technical resources. Without any neglect of the necessary training of new graduates, there are opportunities to strengthen the stability and career patterns for university research, and to increase the effectiveness of interdisciplinary research. Rockefeller University is one example in special areas of the life sciences. Greater attention to this broad need, driven by the very substantial increase in the research management function, should lead to far greater interaction of the university with other parts of the technical enterprise.

NOTES TO CHAPTER 2

1. For statistics see National Science Foundation, "Academic Science/Engineering R&D Funds: Fiscal Year 1982," NSF 84–308, 1984.
2. Ibid.
3. Private conversation with Charles Herzfeld, vice-president for Research, ITT and formerly director of ARPA.
4. James A. Perkins, *The University in Transition* (Princeton, N.J.: Princeton University Press, 1966).

5. Laurence Berlowitz, Richard A. Zdanis, John C. Crowley, and John C. Vaughn, "Instrumentation Needs of Research Universities," *Science 211* (March 6, 1981).

6. "New R&D Centers Will Test University Ties," *Science 227* (January 11, 1985): 150–152.

7. Program announcement for Engineering Research Centers, NSF 84–22, 1984.

8. John Kabler, *The Rockefeller University Story* (New York: Rockefeller University Press, 1970), a brochure.

3 GOVERNMENT ROLES IN U.S. TECHNICAL ACTIVITY

The federal government is probably not where most people would first look to find the source of major technical activity and scientific advances. Nevertheless, it is a principal player in the technical enterprise and potentially the lead player by virtue of its separate roles as performer, funder, and regulator of R&D.

In addition to the increase in government involvement in R&D activity throughout U.S. history, there has been a steady expansion of functions from the relatively passive and indirect to the more active and direct. From the founding of the nation, there has been a series of actions to encourage and strengthen the environment and the infrastructure of science and technology. The basic patent act of 1790 was an early statement of the importance placed on stimulating technical advances, in this case through incentives for technical activity and disclosure. It derived directly from Article 1, Section 8 of the Constitution, which authorizes Congress "to promote the progress of science and useful arts, by securing for limited times to authors and inventors, the exclusive rights to their writings and discoveries." The support for education through the Land Grant College Act in 1862 was in great part motivated by the need for people competent to advance agriculture and the mechanical arts. These were indirect actions, not aimed at a particular technical program or discipline. They were passive, in the sense that the government set conditions, but the initiatives and follow-through remained with private individuals.

As national needs grew and the opportunities for drawing upon science and technology became more evident, the federal government took a more active role in strengthening the technical foundation. The National Bureau of Standards was established in 1901 to facilitate commerce and industry by developing standards for weights and measures. To some extent, it acted as the nation's central research laboratory. The National Institutes of Health, created in 1887, and the Regional Laboratories of the Department of Agriculture are examples of similar active participation. They also represent a more direct function by focusing government actions on selected areas, although quite broad—health and agriculture.

As the twentieth century progressed, government involvement in R&D became active and very direct indeed. The desire to strengthen the technical base of the emerging aircraft industry in the early 1900s led to government initiation in 1915 of the National Advisory Committee for Aeronautics (NACA), the predecessor of the National Aeronautics and Space Administration (NASA). Though driven by military pressures, it represented a major effort concentrated in one technical area. Pressures of war produced direct government programs to produce synthetic rubber and synthetic fuels. And a major by-product of World War II was the substantial government effort to develop nuclear power.

Government participation in R&D has grown from passive and indirect to active and direct. The active and direct government efforts emphasized initially public sector needs—defense, health, agriculture—then extended to areas with mixed features, aviation and nuclear energy, and eventually to concern with private sector needs today.

THE FEDERAL GOVERNMENT'S RESEARCH ROLE TODAY

Actions of the federal government in performing R&D and funding it are difficult to separate. Almost all departments and independent agencies that conduct R&D internally provide some funds for R&D conducted externally. This can be very modest, as in the case of the

Department of Commerce and the Department of Transportation, or extremely substantial, as with the Department of Energy and the Defense Department. The National Science Foundation is the only purely funding agency for R&D.

The range of direct R&D activity by the federal government for the fiscal year 1983 is indicated in Table 3-1. The numbers alone point up some interesting facts. (Keep in mind that all R&D activity in the United States during 1983 totaled $87.7 billion.) The federal government conducted 11.7 percent of total R&D activity and funded 46 percent. Of the total $40.3 billion spent on R&D by the federal government, only $10.2 billion, or 25.3 percent, was spent within government laboratories. The bulk of the funds support R&D in universities, industry, and other institutions. R&D expenditures of the Department of Defense constituted 57.6 percent of all federal R&D expenditures.

The total federal R&D effort can be considered in three categories.

1. *Direct support of the missions assigned to government departments and agencies* such as the Department of Transportation and NASA

2. *Support for basic research*, partly those areas related to missions (Department of Energy and the National Institutes of Health), and partly to strengthen the general reservoir of basic science and engineering (the bulk of funding from the National Science Foundation)

3. *Support for the infrastructure of science and engineering*, such as the science education programs of NSF and the measurement and standards activities of the National Bureau of Standards

The relation between activities and the funding lies more with the source of funds than the performer. For example, the National Bureau of Standards (NBS), one of the finest technical facilities in the world, conducted $141 million of R&D in 1983, of which $55 million was "other agency" money.[1] This latter item represents funds provided by other agencies of the federal government to conduct R&D in support of their interests, primarily because of the competence and facilities available at NBS.

The federal government role in basic research is, again, a mix between performer and funder. While the funding aspect receives most

Table 3-1. R&D Expenditures by Federal Agency, 1983 (*$ millions*).

Department or Agency	Total R&D	Internal	External
Agriculture	$ 848	$ 559	$ 289
Agricultural Research Service	443	401	42
Cooperative State Research Service	232	7	225
Forest Service	108	98	10
Commerce	335	252	83
National Bureau of Standards	95	89	6
National Oceanic and Atmospheric Administration	222	149	73
Defense	22,993	6,401	16,592
Education	112	12	100
Energy	4,537	258	4,279
Health and Human Services	4,353	1,034	3,319
National Institutes of Health	3,789	769	3,020
Housing and Urban Development	32	19	13
Interior	383	274	109
Bureau of Mines	89	58	31
Geological Survey	157	146	11
Fish and Wildlife Services	97	47	50
Labor	20	6	14
Employment and Training	9	1	8
Occupational Safety and Health Administration	5.3	0.4	4.9
State	1.5	0.8	0.7
Transportation	348	76	272
Federal Aviation Administration	126	30	96
Federal Highway Administration	50	2	48
National Highway Traffic Safety	52	8	44
Urban Mass Transportation Administration	57	13	44
Treasury	16	8	8
Other Agencies			
Advisory Committee on Inter-governmental Relations	2.1	2.1	0
Agency for Internal Development	227	21	206
Appalachian Regional Committee	0.2	0	0.2

Table 3-1. continued

Department or Agency	Total R&D	Internal	External
Other Agencies (*continued*)			
Consumer Product Safety Committee	0.5	0.5	0
Environmental Protection Agency	241	93	148
Federal Communications Committee	1.3	1.3	0
Federal Emergency Management Agency	3.3	1.2	2.1
Federal Home Loan Bank Board	3.0	2.8	0.2
Federal Trade Commission	1.4	1.4	0
General Services Administration	0.5	0.5	0
International Trade Commission	4.7	4.7	0
Library of Congress	7.4	6.8	0.6
National Air and Space Administration	2,662	1,134	1,528
National Science Foundation	1,062	131	931
Nuclear Regulatory Committee	207	23	184
Smithsonian Institution	56	56	0
Tennessee Valley Authority	63	43	20
U.S. Arms Control and Disarmament Agency	1.0	0.1	0.9
Veterans Administration	161	159	2
Total	$38,712	$10,582	$28,130

Source: NSF 84-336, Table C-7, p. 26.

attention because the government supports 64 percent of all university R&D (see Table 3-2), basic research conducted within government laboratories is significant. Whether a program is classified as "basic research" is determined in part by the purpose of the program and in part by how it fits into an agency's accounting system, rather than by the particular subject matter and technical objective of the program. This would apply particularly to such large mission agencies as the Departments of Defense and Energy and NASA, where an activity within a particular end-item program is accounted for within that program even though it might be a basic research study. Conversely, an activity of an exploratory nature that is not part of any specific end-item objective may be counted as basic research, particularly if it is performed at a university.

Despite their imprecision, the numbers are interesting for what they reveal about the system of federal government R&D. There are the obvious facts that NSF supports only external basic research,

Table 3-2. Basic Research Expenditures, by Federal Agencies, 1983 (*$ millions*).

Department or Agency	Total	Internal	External
Agriculture	$ 362	$ 239	$ 123
Agricultural Research Service	215	195	20
Cooperative State Research Service	99	3	96
Forest Service	39	34	5
Commerce	19	18	1
National Bureau of Standards	18	17	1
Defense	786	276	510
Education	14	2	12
Energy	768	18	750
Health and Human Services	2,475	532	1,943
National Institute of Health	2,313	449	1,864
Department of Interior	103	84	19
Bureau of Mines	23	16	7
Geological Survey	65	62	3
Justice	3.6	0.1	3.5
Labor	5.3	0.6	4.7
Transportation	0.9	0.2	0.7
Treasury	3.5	3.1	0.4
Other Agencies			
Agency for International Development	4.1	0.6	3.5
Environmental Protection Agency	22.2	8.0	14.2
Federal Trade Commission	1.2	1.2	—
Library of Congress	0.3	0.3	—
NASA	617	305	312
NSF	999	126	873
Smithsonian	56	56	—
TVA	5.6	5.6	—
Veterans Administration	14	14	—
Total	$6,260	$1,690	$4,570

Source: NSF 84-336, Table C-34, p. 65.

while the Smithsonian is a solid performer of basic research solely in-house. A number of agencies are very substantial supporters of external basic research, led by the Department of Health and Human Services through the activities of the National Institutes of Health.

Importantly, basic research takes place within many agencies, and basic research is funded by many agencies. This diffusion of activity is intrinsic to the U.S. system and is very probably a major factor in its past successes and present vigor. The importance of multiple funding is that a researcher has numerous options for obtaining support. Whatever added paperwork or duplication this may produce, there is an enormous compensating benefit in the fact that no agency and no single person has absolute veto over a research concept. Naturally, this depends somewhat on the size and subject of the program, but it is true enough for almost all basic science and engineering proposals of modest size. This element of "marketplace" buying and selling of research is a great source of irritation and strain, but it is simultaneously a relief valve that increases the chance for new concepts to be funded. Further, it exposes the outside researcher to the missions of government agencies and their needs in particular areas of basic science and engineering. In short, it exposes researchers to the technical base required for national objectives.

Of equal importance is the spread of basic research performed within the many government laboratories. Bridges are built between the many federal missions that can absorb new technical advances and the many external sources of basic research. Government laboratories often possess specialized or expensive research facilities that can be an attraction for cooperative research with university and industry scientists. This is clearly true for both the National Bureau of Standards and the National Institutes of Health. In general, the basic research in those laboratories, the Naval Research Laboratory, the Regional Laboratories of the Department of Agriculture, the Bureau of Mines, and many other distinguished government facilities perform the critical functions of providing support for the broad missions of the agencies and developing access to the outside world of science.

Doing much of research related to government interests is the group of national laboratories called the Federally Funded R&D Centers (FFRDCs). These include the major laboratories that serve, and are paid for by, the departments of energy, defense, and several other agencies, but are administered by a university or private organization.

The top ten FFRDC recipients as measured by federal obligations for R&D for 1983 were

1. E. O. Lawrence Livermore National Laboratory (DOE) $603,049,000
2. Sandia National Laboratories (DOE) $523,789,000
3. Los Alamos National Laboratory (DOE) $488,012,000
4. MIT Lincoln Laboratory (Defense)   284,950,000
5. Oak Ridge National Laboratory (DOE)   268,565,000
6. Aerospace Corporation (Defense)   241,532,000
7. Bettis Atomic Power Laboratory (DOE)   241,143,000
8. Jet Propulsion Laboratory (NASA)   226,445,000
9. Knolls Atomic Power Laboratory (DOE)   223,200,000
10. Argonne National Laboratory (DOE)   207,367,000

Two dozen more FFRDCs received amounts ranging from $2 million to $200 million.[2] FFRDCs carried out $900 million of basic research in 1983, or 8.5 percent of the national total. Since the federal government performed $1.65 billion in basic research, the two categories combined accounted for 24 percent of all basic research.

It is instructive to scan the government R&D spectrum for one simple reason. The range of national objectives is reflected to a great extent by the structure of the federal government through its cabinet level departments and its independent agencies. A very great number of these organizational units have developed some R&D activity. Hence, the panorama of government R&D activity represents, in a sense, the identification of the role played by technical change in each of the many aspects of our modern society.

We have no time here for history, but the emergence of R&D facilities within the federal government is itself a history of technical change. It ranges from the older concerns with agriculture and interior, to the newer areas of NASA and the Nuclear Regulatory Commission. It is a fascinating panoply of our civilization.

IMPACT OF THE FEDERAL GOVERNMENT ON THE NATION'S R&D

Beyond the direct government participation in R&D, the government plays a role in determining the conditions under which technical change is generated and used. There are also indirect impacts on the technical enterprise from government R&D activity, largely from defense research. These are both important aspects of the government's role, critical to the health of the overall technical effort, yet far more complex to evaluate than the direct efforts.

This is a very broad area and cannot be dealt with in detail here. I do want to spell out the principal thrusts that should be considered in order to appreciate the cause-and-effect relations between particular actions of the federal government and the R&D efforts of the nonfederal sectors.

This discussion is based in great part on research conducted by the Center for Science and Technology Policy at New York University (NYU). One study concerned the effectiveness of federal support for R&D in the civilian sector. Two publications resulting from this work provide a number of insights and serve as background material for those interested in more detail. The first publication is a scholarly reference work[3] which analyzes the role of government influencing technical change within seven major industries that cover a wide range of technical intensity. The second combines this reference material with an empirical analysis of past and ongoing federal activities to produce an analytical review of the primary conclusions to be drawn.[4]

The federal government has many instruments to draw upon in order to influence the actions of individuals and nongovernmental organizations. These appear to fall broadly into a few categories:

- It can buy products and services.
- It can pass legislation.
- It can establish regulations.
- It can specify taxes.
- It can affect the supply of trained personnel.
- It can provide facilities and services.

In the study of major industries, two questions were raised. First, what have been the sources of technical change? And, second, what, if any, influence did the federal government have on that process for each industry? This second question takes account of the list of actions above.

There is one common thread in the mass of anecdotal data that arises in these analyses. Some actions of the federal government have played an influential part in the technical progress of the major industries studied—semiconductors, aircraft, computers, agriculture, pharmaceuticals, housing, and automobiles. It is very probable that this statement holds for almost every industry.

It is almost a moot point to ask whether the federal government should act to influence technical change in a particular industry. The more realistic question is, What category of federal action will be the most effective and constructive in influencing technical change for that industry?

While federal actions have influenced and will influence technical change in an industry, it is not necessarily true that the actions are taken with that intention. It is therefore important that both Congress and government officials understand and be sensitive to this relationship so that the impact of a particular action on technical change be considered in advance.

Consider a few examples of actions that have influenced specific industries. The most obvious instrument is procurement. This has been a critical driving force in the semiconductor and aircraft industries. Analysis of total semiconductor sales from 1955 to 1977 shows that as they increased from $40 million to $4.6 billion, government purchases went from $15 million to $536 million; the government's share of total purchases decreased from 38 percent to 12 percent as total sales grew.[5] Procurement with the accompanying military specifications stimulated development of both the silicon transistor and the integrated circuit. Similarly in the aircraft industry the years from 1927 to 1933 produced sales of aircraft and engines totaling $185 million, of which $126 million, or 68 percent, was to the government.[6] These purchases plus contracts for delivery of air mail provided the basis for growth and restructuring of the young industry, and thus defined an adequate user market for new technical developments.

Regulation is another powerful instrument of government for influencing technical change. Whereas procurement is almost always a positive force, acting either to induce higher quality performance or to supply a market large enough to absorb the cost of technical developments, regulations can serve either to stimulate or to inhibit technical change. This depends on the industry structure and on whether the regulation serves only to raise costs or provides compensating attractions in the form of a newer or more stable market. For example, building codes appear to have been an inhibiting feature in attracting technical change to the housing industry, largely by the practice of setting standards for specific materials and procedures for installation (and by whom) rather than standards for performance.[7] As the number and stringency of regulations have increased in recent years, the cost and the time of introducing new drugs have increased sharply for the pharmaceutical industry.[8] Thus, the number of "new chemical entities" (NCEs) introduced dropped from an average of about 50 per year in the late 1950s to about 17 in the late 1970s. The cost of developing a single NCE is estimated to have risen from about $1 million to $2 million in the early 1960s to roughly $24 million in the early 1980s.

Conversely, regulations regarding emissions and fuel efficiency have clearly forced technical change in the automobile industry. The argument here is whether this process has been more costly than other approaches, but technical change has indeed resulted.[9] Airline regulation provided a market for newer and more sophisticated equipment, since the airlines were competing on service rather than rates.[10]

Technical change in agriculture has been dramatic, and results from a combination of factors. Advances by the food machinery and fertilizer industries are obviously major contributors. Government actions in different forms, however, have had substantial influence.[11] The Land Grant College System served to provide trained manpower and to focus research interests at these universities. Agricultural Experiment Stations provided by the U.S. Department of Agriculture and an additional 52 State Agricultural Experiment Stations provide a broad base of R&D and services which can transfer technical advances to the users. This system is strengthened by the Federal Extension Service, developed just before 1925.

Probably the most influential, yet most complex, action of government lies in changing the financial conditions for conducting and exploiting R&D. Fundamental to any evaluation is that the decision to conduct a technical program in the private sector is based on the ultimate economic value to be gained by converting the results into use. Any action that lowers R&D costs or increases after-tax returns will increase the probability that an R&D program will be initiated.

Whether this cause-and-effect relation can be pursued by design is a difficult question. If a change in taxes benefits R&D, it very likely benefits other investments as well. Tax incentives for increased R&D expenditures help established firms, but do little for new companies. Rapid depreciation of equipment can be a stimulus to purchasing new scientific instruments, and can improve the return on R&D in the process industries, but will not be sufficient alone to compensate for declining markets.

The fact is that R&D constitutes just one part of a complex economic system. Financial actions, whether intended to promote R&D specifically or to stimulate economic growth generally, are absorbed into the total economic environment, and the ultimate impact on technical change is difficult to predict or control. To take just one example, R&D is a lengthy activity from initiation through economic use. Its benefits are therefore highly sensitive to prevailing interest rates. As another example, the ability to sell products resulting from R&D is strongly affected by currency exchange rates, which can give competing foreign products a considerable price advantage.

Thus, while governmental actions can indeed influence the financial conditions for conducting or converting R&D, we must be careful not to consider those actions in isolation. Other perturbations of the financial system can neutralize or even prevent an intended action, and other factors may be far more influential than a modest attempt to improve the financial odds.

I have no intention of reviewing the broad impact of government on innovation. Perhaps the most thorough study in recent years was conducted in the late 1970s under Jordan Baruch, then Assistant Secretary of Commerce.[12] Roughly twelve task forces dealt thoroughly with such issues as the effect of regulations, antitrust laws, tax incentives, and the like.

My intent here was simply to indicate the pervasiveness of government actions in influencing technical change, whether by design or as a by-product, through a great many mechanisms other than the direct conduct or funding of R&D. There is, however, one area that should be looked at separately. This is the impact on nongovernmental technical activities resulting from governmental technical activities.

The most important and most interesting issue based on both dollars and philosophy is the impact of defense research on the rest of the technical enterprise. There are a number of different aspects to this issue which are valuable to consider because they illustrate the interactions that can occur between the government sector and other technical activities. Perhaps the simplest way to identify these items of particular concern is to pose some questions.

- Does the allocation of technical resources to defense R&D diminish the technical strength and international competitiveness of the civilian sector?

- Can the results of defense R&D activity be useful to the civilian sector?

- Do the demands of defense R&D for technical personnel and scientific equipment create scarcities and raise costs which result in a decreased civilian R&D effort?

While the questions can be stated separately, the answers cannot. Any discussion of one runs into the others. Let us consider the situation created by defense R&D and the assumptions about the system generally.

One fundamental line of reasoning is that industrial research provides an effective conversion of technical advances into economic use because it is integrated within an overall system of manufacturing and marketing. A given amount of defense research will not, in general, yield the same economic results for the industrial sector across the board that would result from the same amount of research in industry. This is correct, though we would have to make exceptions for the aircraft industry and the early days of semiconductors and computers, for which defense R&D supported the technical foundations. These are not trivial exceptions, particularly with respect to international trade.

Thus, if our primary objective is the development of economically useful products, processes, and services, and if there is a fixed amount of technical resources (people, equipment, funds), then any diversion of these resources to defense research detracts from the maximum output of new goods and services. We pursue defense R&D because national security is critical, but we thereby lower our economic potential.

That is the simplified argument. The system, however, is much more complex, and the reasoning must be examined carefully. I had accepted and pursued this reasoning for much of my own professional career. There is, however, reason to believe that we have probably underestimated the overall positive impact of defense research on the technical system generally. Certainly my colleagues in Europe, both in government and industry, have long pointed to that as a major factor in American technological superiority and the consequent increase in industrial strength and international competitiveness of the United States. I used to argue with them that our defense research, however necessary, took away resources from civilian research. At the very least, we should not consider the percentage of gross national product (GNP) devoted to R&D to have the same impact on our industrial strength as would be true for, say, Japan and West Germany, since defense R&D accounted for so much of the U.S. expenditures.

In fact, my friends in Europe were perhaps more correct in their thinking than I and most of my colleagues in U.S. industrial research. The considerations to follow are along these lines:

First, there is in fact very little fallout into direct civilian commercial use of the devices, materials, or components emerging from defense R&D, certainly negligibly so in relation to the cost. Such spin-offs should not be the basis for evaluating benefits. There have indeed been items of great value, and it is very probably worth added investment to identify and extract even more, but no serious argument can be advanced to justify the substantial funding of defense R&D in terms of the very insubstantial economic value from direct transfer of outputs to civilian use.

Second, the particular industries that have been direct beneficiaries of defense R&D have produced very broad impacts. The aircraft industry became an important contributor to the U.S. export base and the dominant force in world aviation. Even more broadly, the electronics industry emerged as the technical leader in semiconductors, computers, and telecommunications. Not only are these industries important domestically and in exports, but they contribute to the upgrading and increased productivity of a great many other industry sectors in which technical change is advancing based upon the revolution in electronics.

Third, defense R&D preserved and expanded the university research system of the United States during its phenomenal growth in the two decades from 1950 to 1970. The deliberate actions toward this objective of the Office of Naval Research (ONR) immediately after World War II bought time for the National Science Foundation to play a more substantial role. Increasing defense support for R&D in almost every area of methematics, physical sciences, and engineering at universities has been a major factor in the increased technical sophistication built into graduate training for at least 20 years. This has provided a generation of advanced science and engineering graduates which has brought to every sector, including industry, the results of that training. One indirect impact of defense R&D, through its university research support, has been to raise the level of science and engineering in all sectors of activity in the United States.

Fourth, the concern that the demands of defense R&D can create scarcities of technical personnel or equipment implies that this is a "zero-sum" situation—that there is a fixed amount of people and equipment, so that their use for one function means there are that many less available for another. To begin with, our system simply does not work that way. If we did not use a certain number of physicists or electrical engineers in defense R&D, there is no reason to believe that industry would add them to their own R&D programs. Much more fundamental, however, is that this is *not* a zero-sum situation. There is every indication that the growth of defense R&D in the 1950s and 1960s with its accompanying support for university research attracted many people to careers in science and engineering. The thrusts in new technologies were glamorous, the well-advertised needs of defense R&D indicated that jobs were available, and the funding of university research provided financial support for graduate students. The combination of support today and glamorous jobs tomorrow undoubtedly resulted in increasing the number of scientists and engineers with graduate degrees.

The defense R&D effort has strengthened our technical base generally and our industrial research effort in particular. Could this have been accomplished more effectively and more directly by industry instead of government? Of course, but only if the particular industries most affected were to increase their R&D expenditures to a point where the companies might not be competitive. Could the civilian sector have been helped more effectively if government R&D programs were intended for that purpose? Of course not, unless we believe that government officials are better suited to select, conduct, and transfer those R&D programs most appropriate for industrial growth.

This is not of historical interest only. Today, defense R&D is increasing as a share of federal R&D expenditures. The fiscal 1985 budget shows defense R&D increasing from $27.4 billion in fiscal 1984 to $34.9 billion, or 27.5 percent, while the total federal R&D budget increases from $44.4 billion to $52.7 billion, or 18.7 percent.[13] This means that defense R&D rises from 61.8 percent of all federal R&D funds to 66.3 percent. Obviously, federal expenditures for other areas related to civilian sector R&D increase only slightly or are cut back.

Objections are being raised to this trend by those who are convinced that objectives for the civilian sector are endangered by further increases in defense research.[14] My point is simply this: there may well be sound philosophical or financial reasons for objecting to further increases in defense research, but concern for the civilian sector and our international competitiveness is not one of them.

I am stating only that defense R&D has been valuable to our technical base and our industrial strength. I sincerely hope that this will not be carelessly twisted into the converse statement, namely that, in order to improve our technical base and our industrial strength, we *must* pursue defense R&D. Defense R&D is not at all a necessary means to broaden our technical efforts. The characteristics that have been of value to the technical enterprise could in principle have arisen from other actions which incorporate:

- A major national objective, extending over a period of years, was supported strongly by broad constituencies.

- Progress required technical advances across a wide spectrum of science and engineering.

- Activities involved the nongovernmental sectors, particularly the universities.

These characteristics are not unique. They are common to other branches of the federal government. The programs of the National Institutes of Health laid the foundation for the blossoming of biotechnology, which is steadily affecting the manufacturing processes in the chemicals, food, pharmaceutical, and energy industries. The R&D expenditures are only about 20 percent of defense R&D, and the activities are focused on the life sciences, but the breadth of impact on our technical base is impressive. Another large national program is obviously in space. This does have wide interests represented by defense, and is high technology in the most literal sense. However, its R&D expenditures are roughly 15 percent of defense R&D, and its continuity and growth do not have the strong broad constituencies of defense, despite the challenging nature of the space program.

The argument, then, is not that defense R&D is necessary to strengthen the technical base of the civilian sector. It is not even that large government R&D programs are necessary for that purpose. Rather, it is that a large government R&D program that has continuity, that draws upon a wide range of science and engineering, that provides important roles for university and industry, and that focuses this range of technical activities upon well-defined objectives of the public sector will have broad value for the overall base of science and technology and thus for our industrial strength generally.

It does not follow that *any* large-scale federal support for R&D will produce proportionate benefits. It certainly does not follow that any large-scale federal support *intended* to strengthen the civilian sector will provide reasonable value, and there are arguments to demonstrate the possibility of doing active harm, depending on the choice of programs and means for exploitation.

There is a slightly amusing final note concerning the benefits from defense R&D for our international competitiveness. The cover story in *Business Week* of March 14, 1983, focused on "Rearming Japan." It addressed the growing pressure from the U.S. government for Japan to increase its spending on defense and on defense R&D in order to share the burden of a common effort for cooperative national security. However, a portion of the overall story deals with the concern that "many U.S. executives are growing increasingly apprehensive about the business fall-out should Tokyo agree." While part of this concern refers to competition in arms sales, a more important part is that "the Japanese may be able to apply this defense technology to commercial markets ranging from civilian aircraft to communications satellites." We seem quite prepared to give Japan credit for assets related to defense research that we downplay in our own activity.

THE ROLE OF OTHER NATIONS' GOVERNMENTS IN R&D

In all of the developed countries, and an increasing number of developing countries, government engages in the activities and functions related to R&D that we have discussed in this chapter. A rough comparison of the magnitude of this activity and its relationship to the private sector is indicated in Table 3–3, based upon data from the OECD for the year 1979. (The table does not indicate government funding of university R&D.)

The data from the OECD is not as detailed as the data now systematically prepared for U.S. expenditures by the NSF. Nonetheless, observations about the role of government can be made. For the larger and the medium-size countries, industry funds from 40 percent to 65 percent of all R&D activity. This drops off in the smaller

Table 3-3.   R&D Expenditures in OECD, 1979.

	Total R&D ($ billions)	Univ. R&D (%)	Govt. R&D (%)	Industry R&D (%)	Percentage of Industry R&D Funded by Govt.
United States[a]	$56.56	14.7	17.9	67.5	32.7
Japan	18.19	27.4	14.4	58.2	1.6
Germany	12.53	16.0	14.9	69.1	20.7
France	7.96	15.5	25.0	59.5	28.7
United Kingdom (1978)	7.96	11.4	24.4	64.2	37.2
Italy	3.09	17.8	23.9	58.3	6.1
Sweden	1.61	20.5	13.0	66.5	14.0
Belgium	1.07	20.6	9.3	70.1	6.7
Norway	0.52	30.8	19.2	50.0	26.9
Yugoslavia	0.59	18.6	28.8	52.5	22.6
Ireland	0.12	18.1	44.8	37.1	14.0
Portugal (1978)	0.09	14.3	72.5	13.2	8.3

Source: OECD Science and Technology Indicators, OECD, 1984.

Note: Data is combined from several tables of the OECD report, and it is assumed that the funds spent on R&D conducted by the government is what remains after subtracting the R&D spent in industry and universities from the total national R&D expenditures.

a. The figure given by NSF for 1979 is $54.98.

countries. All governments fund a significant portion of R&D performed in industry, much of it related to national security. The dramatic exception to this is Japan. Possibly contradicting much popular thought in the United States, the Japanese government funds just 1.6 percent (in 1979) of R&D performed by industry. This is related to the minimum efforts in defense R&D, but there is clearly no significant direct funding of commercially oriented R&D either. The government role in strengthening Japan's international competitiveness is not pursued with direct R&D funding to the companies though funds do flow through certain collective research associations.

Support for R&D in the universities comes almost entirely from public sources in most of the countries listed. The OECD report indicates that Japan provides only 59 percent of such support from the government and the United Kingdom 80 percent, while the other countries range from about 90 percent to 100 percent.

The patterns with respect to R&D funding are therefore not that different throughout the developed nations. Government provides for the infrastructure with its support for university research, and it accounts for the direct needs of the public sector through the R&D within government laboratories. Government funding for industry R&D is highest proportionally for the larger countries, related to defense programs. Government plays a larger role in the smaller countries, presumably because the minimum resources necessary for organized R&D are available largely through government.

Numbers, of course, are not the whole story. The various mechanisms are available to government to influence decision-making on the direction and magnitude of R&D activities in industry. The United States may rely somewhat more on flexible financial incentives and multiple decision points within industry, while other countries may use procurement more widely and more purposefully. These, plus related regulatory activities, are often tailored to such specific objectives as energy, health, and environment with resulting impact on R&D in the affected industries. Other countries are much less bashful about identifying and advocating specific industrial policies, then gearing both direct government R&D and indirect mechanisms to their implementation.

Another factor of major importance is that industry in other countries is not always "private" and, where it is, can have interaction with government that is far different from that of the United States. Obviously, these circumstances result in substantial differences in the role of government regarding R&D activity. The OECD data all refers to "BERD"—that is, R&D in the business enterprise sector, which includes both public and private companies. Many companies in the energy sector are wholly owned by governments, and governments own large equity positions in many major corporations throughout the OECD membership.

When these public–private relationships are considered in more detail, it is clear that government involvement in, and influence upon, R&D programs within industry are more intensive in countries outside the United States than are indicated by the expenditures shown in the table. To begin with, the numbers themselves, if adjusted to credit government with R&D funding for the publicly owned or dominated business enterprise, would show substantial increases in the proportion of national R&D funded by government.

This alone is not very critical. Senior technical executives in European firms which are government owned or dominated do not act much differently than senior technical executives in U.S. corporations with regard to internal management of R&D, converting technology to use, or planning the optimum technical support for the current objectives and future growth of the enterprise.

The government role in other countries becomes relatively more important in those aspects of the technical enterprise for which linkages with external technical elements are valuable to the conduct and direction of technical growth. These linkages will be discussed in more detail later on, but they have been of great importance within the OECD membership.

The most obvious link is between R&D programs of the government and industry. This follows directly from the observation above that most other countries have a defined or implicit industrial policy with regard to encouraging particular industries or adopting objectives that point to particular areas, for example to emphasize export of high-value-added, high-technology products. When such policies are sufficiently identified, the directions of technical activity follow logically. When a country states such policies explicitly to favor particular industries, a national science and technology policy falls easily into place.

Hence, government R&D activities and those of the enterprises are often supportive of each other, partly by design and partly through cooperation. Such cooperation is a source of great current activity in microelectronics and telecommunications in Japan, France, Germany and the United Kingdom, where those areas have become key elements in national industrial strategy.

Similar pressures influence joint ventures, mergers, and acquisitions between domestic firms within a country or between a national enterprise and a foreign company. Government considerations with regard to a desired technical strength have been involved directly in France (reducing Honeywell's equity in CII–Bull, and bringing St. Gobain into the field), in the United Kingdom (encouraging Thorn–EMI to take over Inmos), and in many other countries and examples. The relationships between enterprises, domestic and foreign, are initiated, encouraged, or inhibited based increasingly on the desire to strengthen selected technical directions and competencies.

One area where there seems to be little difference between the United States and other countries as far as a government role is concerned is in the relationships between university and industry. All governments encourage the optimum use of the total technical resources of a country to meet national objectives. Those factors which bring university and industry to work cooperatively or to keep them apart are the same everywhere. A major financial difference, of course, is that in most other countries the university system is a government responsibility, so that funding for salaries and for research comes directly from the government, national or regional. Thus, while added industrial funding is always desirable for many reasons, the financial pressures are different. Nevertheless, the one major international conference held on the subject of university–industry interactions (in Stockholm, March 1983) brought out clearly that the differences between industry sectors on the interactions were greater than the differences between countries.[15]

This discussion of government's role in R&D has focused on the developed countries within the OECD, which contain the private sector and the growing resources devoted to industrial research. This is the principal focus of the present book. However, from a global view, we should consider the Soviet Union, where there is no separation between public and private sectors.

The conduct of R&D in the Soviet Union is subject matter for a separate book. Several references can be suggested for those concerned with it.[16] I will not even attempt to summarize this area, partly because it is so large and specialized, but also because it is not relevant to the main thrust of my arguments concerning the process of technical change.

The Soviet system is simply not effective in the over-all process by which we integrate the planning and conduct of R&D with the conversion to economic use in the civilian sector. The USSR pours massive resources into R&D, roughly 1.5 million professional researchers compared to 750,000 in the United States in 1983.[17] It generates science and technology, and it is obviously successful in converting results to use in military applications where the user can define specifications and control manufacturing processes. These conditions do not apply in the civilian sector. There are indications that there are two separate levels of technical activity in the USSR — military and civilian — with different quality of facilities and procedures.[18]

We tend to compare the relative international competitiveness of countries by such indexes of technical intensity as the proportion of gross national product devoted to R&D or the number of scientists and engineers per 10,000 employed or the number of technical graduates as a percent of total graduates. The USSR is a leader in all of these categories, yet we do not rate it highly as a commercial competitor. This should remind us to use such indexes with caution, since they are no measure at all of the technical system, the interactions of the components within that system, and the effectiveness of that system.

NOTES TO CHAPTER 3

1. Discussion with Edward Brady, Associate Director, National Bureau of Standards, spring 1985.
2. See complete list in Appendix B, Federally Funded Research and Development Centers, Fiscal Years 1983–85, in Federal Funds for R&D: Fiscal Years 1983, 1984, and 1985, NSF 84–336, 1984.
3. Richard R. Nelson (ed.), *Government and Technical Progress* (New York: Pergamon Press, 1982).
4. Herbert I. Fusfeld, Richard N. Langlois, and Richard R. Nelson, "The Changing Tide," Center for Science and Technology Policy, Graduate School of Business Administration, New York University November 1, 1981.
5. Richard C. Levin, "Semiconductor Industry," in Nelson, *Government and Technical Progress.*
6. David C. Mowery, and Nathan Rosenberg, "Commercial Aircraft Industry," ibid.
7. John M. Quigley, "Residential Construction Industry," ibid.
8. Henry G. Grabowski and John M. Vernon, "Pharmaceutical Industry," ibid.
9. Lawrence J. White, "Motor Vehicle Industry," ibid.
10. Mowery and Rosenberg, "Commercial Aircraft Industry," ibid.
11. R. E. Evenson, "Agriculture," ibid.
12. *Domestic Policy Review of Industrial Innovation*, U.S. Department of Commerce, Office of Assistant Secretary for Productivity, Technology and Innovation, May 1978.
13. "Research & Development, FY 1985," AAAS Report IX, American Association for the Advancement of Science, 1984.
14. Franklin Long, "Federal R&D Budget: Guns versus Butter," Editorial in *Science* (March 16, 1984).
15. H.I. Fusfeld and C.S. Haklisch, *University Industry Research Interactions* (New York: Pergamon Press, 1984).

16. See bibliographies contained in *Soviet Science and Technology*, edited by J.R. Thomas and U.M. Kruse–Vaucienne (Washington, D.C.: George Washington University, 1977).
17. "International Science and Technology Data Update," NSF, January 1985.
18. David Holloway, "Soviet Military R&D: Managing the Research Production Cycle," in Thomas and Kruse–Vaucienne, *Soviet Science and Technology.*

4 INDUSTRY
Now the World Player

The emergence of industrial research is a phenomenon of the twentieth century. It is such an accepted part of civilization within the developed countries that most of us hardly think about it at all. Only when we consider the struggles of developing countries to establish a modern industrial base or reflect upon the nature of industry in earlier eras do we begin to sense that there has been a fundamental change in industry—a change in the nature of industry, not just its size and the products manufactured.

Trade and industry over the centuries have always absorbed advances in technical knowledge, improved upon them in use, and encouraged inventors to make our lives more comfortable and extend our control over nature. The only change at the opening of the twentieth century was to conduct some of these activities within the industrial organization itself.

The change may seem small, but ah, what a change that was! It directed investigation, accelerated conversion, provided the support structure, and stabilized funding. Most of all, it changed thinking and relationships within the technical community. In doing so, it changed, and is changing, the strategic basis for industrial growth.

The next chapter shows how the evolution of industrial research was a reaction to the increasing cost and complexity of the technical process. This chapter examines the nature and scope of industrial research and how it propels the entire technical enterprise.

Technical development occurred in earlier centuries as manufacturing and mining enterprises adopted mechanical advances, repaired and maintained tools and equipment, improved products and processes, and performed quality control, however elementary. The integration of science and technology became more sophisticated by the late nineteenth century with the growth of the electrical and chemical industries, which required more links with new scientific phenomena than the mechanical systems upon which much of the Industrial Revolution was based.

The earliest industrial research laboratories were very humble.

> On the banks of the Erie Canal in December 1900, GE's chief consulting engineer, Charles Proteus Steinmetz, had led out of his boarding house and into a barn in the backyard of his house an instructor in chemistry from MIT named Willis R. Whitney. The barn became GE's first laboratory, and Steinmetz installed Whitney there as GE's first director of research.[1]

From these simple beginnings, industrial research has expanded at a staggering pace, accelerated sharply by both World Wars I and II. The objectives of industrial research are stated simply enough:

- To strengthen present products and processes
- To develop new products and processes that will expand present business
- To provide the basis for new business

Notice that these objectives do not distinguish basic research from applied research or engineering. Nor do they mention disciplines, such as physics or chemistry or electrical engineering. The objectives are stated in terms of functions to be performed. Each company can identify specific missions that relate to its particular products or business interests, whether solid state controls or oil refining. These missions then determine what know-how will be needed to pursue the goals. Thus industrial research is, by definition, interdisciplinary. The particular mix of physics, biologists, chemical engineers, and others that make up an industrial research laboratory follows from the missions the parent company has established for that laboratory.

In a large company, especially the multinationals, industrial research occurs in numerous facilities. This is because a large company is rarely a single product or a single business operation. Each broad product area often requires a technical group dedicated to strengthening present products and processes and contributing to expansion of that business in the near term, say two years ahead. The corporation as a whole must make sure to devote enough effort to long-range R&D as well if it is to continue to thrive. It may even invest in developing a base for new business growth not necessarily related to present businesses.

The conventional approach corporations use to accommodate these different objectives is to establish a central R&D facility as well as an R&D organization in each operating organization. The central R&D organization is part of the general management of the parent corporation. The senior technical executive is typically vice president of R&D or technology for the corporation. As an officer of that corporation, that executive is usually responsible for linking central R&D and the R&D activities of the operating units, and serves as the senior staff adviser for science and technology to top management.

The division of responsibilities among these multiple technical organizations within a large corporation must be kept in mind by anyone who wishes to understand the nature of industrial research or to influence its direction or output. For example, consider the question of whether government-funded R&D conducted by a large corporation influences or benefits other technical activities of that corporation. This is not too different from asking whether an R&D program at one location influences or benefits other technical activities throughout the corporation. It is obvious that the answer must take into account many questions of communication, management of resources, mobility of personnel, and funds transfer within the company. There is no automatic yes or no to the question.

Another example is the ongoing debate about whether tax incentives stimulate R&D. Answering this question requires a careful analysis of the accounting procedures within each corporation to determine whether the benefits of the lower effective R&D costs are available to the head of an R&D facility within the corporation in a way that affects the planning for future programs.

The great many research projects supported within a corporation require different amounts of time and money for completion. Thus, the overall investment of a corporation in its technical activities is similar to a broad investment portfolio: there is high probability of modest returns from short-term activities related to present products and businesses and lower probability of very high returns from long-term activities that would change present businesses drastically or lead to new business opportunities.

It is this range of probabilities that justifies the heavy investment by industry in R&D. A single project which has the promise of high return may represent high risk, but the total activity does not. Projects that may represent glamorous breakthroughs are normally a small part of total activity. The bulk of corporate R&D is devoted to improving present products and processes and providing technical services. The typical division among the principal functions performed by industrial R&D activities is shown in Table 4-1.

Table 4-1. Distribution of Effort in Industrial R&D (%).

By Function	1970	1981	By Objective	
Basic research	3.3	3.2	Wholly new businesses	0-10
Applied research	19.0	20.7	Present businesses New products and processes	15-25
			Improved products and processes	50-65
Development	77.7	76.2	Technical services to operating divisions	5-15

Sources:
By function: Table B-27 in "R&D in Industry, 1981," NSF 83-325.
By objective: discussion with W. Abel, Chairman, Committee on Definitions of Research, Industrial Research Institute (1976).

There is one important guideline for investors wishing to profit from technical developments, regions desiring to use technology as a basis for economic growth, and policymakers seeking to stimulate technical activity. Stated simply: *Industrial research is successful when the technical activity is an integral part of the company.* Any separation of R&D from current operations and future plans of the corporation lowers the probability of success. I am not referring to the physical separation of the R&D facility, which can have advantages for long-term programs, but to the planning and conduct of R&D independent of considerations for manufacture, marketing, and financing of the output.

Three characteristics are most critical to the success of industrial research:

1. The range of programs and functions must be large enough to ensure that much of the effort adds value to the ongoing businesses of the corporation, which can translate the output into profits through greater sales or lower costs.

2. The technical programs selected must fit the corporate capabilities and growth strategy, so R&D has a high probability of being converted to use successfully.

3. A number of research projects in progress at a large industrial laboratory must be highly likely to succeed and useful outputs should emerge at different intervals.

Thus, industrial research contains the critical elements of *balance* within the corporation, *mix* of functions and *quantity* of projects. Essential to all three elements is that technical activity not be planned and conducted in intellectual or organizational isolation, but with respect to a very particular corporation. A successful program at Eastman Kodak is unlikely to be appropriate for Alcoa.

Developments that prove valuable to other corporations in the same or a different industry may eventually be a source of secondary income to their original sponsor. Licenses may be granted to the other corporations so that a stream of royalties will flow to the sponsoring corporation. However, I am not aware of any instance where a major industrial research facility receives enough income from licensing to justify its original creation and continuing operation. If that were the case, the financial community would invest in laboratories, and they do not.

This is not a trivial point. There are trends today in new mechanisms for financing R&D projects, in early venture capital, and in regional economic development that encourage R&D activity not integrated within the organization which will be responsible for its conversion and marketing. The intent is to exploit the results by licensing to other corporations or by establishing new ventures. The concept that one can conduct independent R&D as a profitable activity based upon the output is not impossible, as witness the original Edison Research Laboratory, but it is not industrial research. It lacks the statistical protection of quantity, of functional mix, and of higher probability for effective transfer provided by the parent corporation.

SCALE OF INDUSTRY RESEARCH

The most obvious aspect of industrial research in the United States is its size. In 1983 industry conducted $64.6 billion and funded $44.4 billion of R&D activity. This represented 74 percent and 51 percent, respectively, of the total national effort. Of the 750,000 scientists and engineers engaged in R&D in 1983 556,000, or 74 percent, worked in industry.[2]

Once we look beyond the overall industry figures, the next most obvious feature is the difference in the level of R&D for the separate industry sectors. This quickly raises two questions that are critical to understanding the role and impact of industrial research:

1. What portion of R&D conducted within an industry sector is funded by government?

2. What is the R&D intensity of an industry sector—that is, what percentage of sales is spent on R&D?

Consider the first question. Government-funded R&D is pursued under wholly different criteria than industry-funded R&D. Government-funded R&D is a product. Company-funded R&D is an investment that can only provide a return when the results are converted into products, processes, or services. There are indirect benefits from government-funded R&D. Intelligent use of government-funded R&D can indeed leverage the use of company-funded R&D, particularly in advances within the aircraft and electronics industries. Nevertheless, any comparative analysis of the investment in industrial research as a mechanism for corporate growth should focus on industry-funded R&D, with perhaps some adjustment to consider the impact of substantial government funding in a few industry sectors.

The second factor of R&D intensity or technological intensity of an industry is described most commonly by the percentage of sales that is devoted to R&D. Since this is a ratio, a few anomalies show up. The oil and automotive industries are highly advanced technologically, greatly affected by technical change, and responsible for very large R&D efforts. However, because their gross revenues are so large, their R&D expenditures as a percentage of sales are low.

R&D intensity, the subject of the second critical question, is primarily a description of an industry's structure, not a measure of "good" or "bad" performance, as the previous comment on the oil and automobile industries should have made clear. The technical base necessary to maintain an industry's product lines, the rate at which new products are introduced, the need to upgrade processes all lead to some portion of the sales dollar being allocated to R&D. Any given company in an industry will deviate from the industry average, but not by too much or for too long a period. If a company spends consistently much lower than the average, any short-term cost advantage will be offset by fewer new products or lower quality. A company that spends consistently higher may find that it has a lower profit margin. If the higher R&D intensity is successful in new products and market share, it may force the industry average higher.

Some of the prominent data on industrial R&D are shown in Table 4-2 for industry groupings used by the National Science Foundation, based upon the Standard Industrial Code (SIC).

The table reveals the sharp differences among industry sectors regarding the conduct of R&D, significant when contrasted with the "average" figures normally used to describe industrial research. Among the more pertinent characteristics are these:

- Government-funded R&D in industry is highly concentrated. Just two sectors—aircraft and missiles plus electrical equipment— receive 76 percent of all such funds.

- Industrial research overall is highly concentrated. Three sectors— aircraft and missiles, electrical equipment, and machinery— account for 56 percent of the total R&D conducted by industry; three sectors—electrical equipment, machinery, and chemicals— account for 50 percent of all R&D funded by industry.

- R&D intensity based on industry funding is highest for office equipment, computers, instruments, and drugs and lowest for food, textiles, paper, and metals. These levels do not correlate with the level of sales, but with the nature of the different sectors.

The sheer magnitude of U.S. industrial research is one of its outstanding features. It should be held in mind constantly for reference when discussions occur about increases by other countries in the *percentage* of GNP going to the respective national R&D efforts. It should also provide a continuing perspective when programs are

Table 4-2. R&D Characteristics of Industry Sectors, 1981 ($ millions).

Industry	Funded by				R&D as a % of Sales	
	Total R&D	Govt.	Industry	% Govt.	Total	Industry-Funded
All industry	$51,810	$16,382	$35,428	31.6	3.1	2.1
Aircraft and missiles	11,968	8,528	3,440	72.3	16.0	4.4
Electrical equipment	10,329	3,920	6,409	38.5	6.8	4.2
Primary metals	878	176	702	27.8	0.9	0.7
Professional and scientific equipment	3,614	637	2,978	16.0	8.1	6.8
Motor vehicles	4,806	587	4,219	12.1	4.5	4.0
Machinery (including office equipment)	6,818	694	6,124	10.9	4.9	4.4
Fabricated metal products	624	80	545	10.0	1.4	1.3
Chemicals and allied products	5,625	421	5,205	6.6	3.6	3.4
Lumber and wood products	161	0	161	0.0	0.8	0.8
Paper and allied products	566	0	566	0.0	1.1	1.1

Source: Tables 38 and 42 in "National Patterns of Science and Technology Resources, 1984," NSF 84-311, 1984.

announced by other governments or a group of governments to support particular areas of research in accord with their industrial policies.

In the most active area in recent years, microelectronics, programs have been initiated by France, West Germany, and the United Kingdom to support R&D with hundreds of millions of dollars by the separate governments. The European Economic Community is funding approximately $750 million to match similar private funds in ESPRIT, a cooperative program to strengthen research in information technology. The U.S. government has nothing of such magnitude outside of the Department of Defense. But contrast this with our *private* industrial expenditures of many billions of dollars of R&D in semiconductors and computers. The R&D expenditures of IBM alone in 1985 is estimated at $3.5 billion.[3]

Thus, when we consider the conduct of R&D intended for the civilian sector for economic growth, we cannot consider U.S. government actions in comparison with those of other governments. American industrial research possesses the critical mass necessary for effective growth in almost all industry sectors, a situation that does not hold in all other developed countries.

One important aspect of industrial research deserves comment. Its primary strength derives, of course, from its interaction with the resources and growth of the individual corporation. Nevertheless, the outputs from this strong technical base serve to strengthen all industry, indeed all science and technology, in two broad categories:

1. Specific technical advances in areas such as instruments, electronics, and materials are absorbed throughout all industry and act to increase productivity across the board.

2. Contributions to basic science and technology increase the world level of knowledge and know-how.

The contribution of R&D to industrial productivity is obvious in principle, based upon development of lower cost processes and higher value-added products. The contribution of R&D in one industry to productivity in another is perhaps equally obvious when we think about it, both through products such as instruments and computers, as well as through broadening the technical base. To develop quantitative measures is quite another problem. Major work in this field is being pursued by the New York University economist, Wassily Leontief, who received the Nobel Prize in Economics for his general work in input–output theory.[4] Applying this methodology to R&D may add to our understanding of the value added by the industrial technical base.

The nature and extent of basic research in industry deserves attention because it affects the entire technical community, and its role in industry is perhaps not well understood. It is not an objective of industry to do basic research or to conduct R&D. However, R&D has proven to be a necessary and cost-effective activity for preserving the health of present businesses and providing options for growth, which are objectives of any corporation. By the same rationale, allocating some proportion of the R&D budget to basic research can prove worthwhile under the following conditions:

1. If basic research can save time and cost by guiding empirical design engineering
2. If it leads to alternative approaches in a broad mission-oriented program through new materials, components, or systems
3. If it provides options for wholly new products, processes, or services
4. If it serves to develop liaison with the worldwide scientific community to strengthen access to knowledge and people

Fulfilling these conditions is quite different for different industries, for different companies, for different programs. One direct economic inhibition is that the results of basic research are not allocable completely to the sponsoring firm, but are available to all who can use them. There can be delays built into this process and specific know-how can be kept confidential, but basic technical advances will diffuse into the scientific community.

Despite the fact that one company's funding produces some benefits for all, there are incentives to pursue basic research. Industrial basic research is (or ought to be!) directed basic research. That is, the company has a good reason for supporting a particular area of research, perhaps the particular project. Thus, the transfer process should be effective with minimum time delay. A company with a very broad technical base can have a high probability of finding some application for a well-chosen program in basic research. Finally, a company which has a large market share in a line of business can derive sufficient value from a related area of basic research that it is a sensible activity even if others derive benefit.

Table 4-3. Distribution of Basic Research in Industry, 1981 (*$ millions*).

Industry	Total R&D	Basic Research	Basic Research as % of Total R&D	Basic Research as % of Total Industry Basic Research
Total	$51,830	$1,641	3.2	—
Aircraft and missiles	11,702	128	1.1	7.8
Electrical equipment	10,466	279	2.7	17.0
Machinery (including office equipment)	6,800	128	1.9	7.8
Chemicals and allied products	5,325	539	10.1	32.8
Motor vehicles	4,929	21[a]	0.4[a]	1.3
Professional and scientific equipment	3,685	40	1.11	2.4
Primary metals	889	46	5.2	2.8
Food and kindred products	719	27	3.8	1.6
Fabricated metal products	638	8	1.3	0.5
Paper and allied products	562	32	5.7	2.0
Lumber and wood products	167	1	0.6	0.06

a. Company funded.
Source: Tables B-28 to B-32 in National Science Foundation, "R&D in Industry, 1981," NSF 83-325, 1983.

Nevertheless, there are indeed reasons why not every company or every industry has equal justification to support basic research. This unequal distribution is clearly apparent in Table 4-3 giving industry performance for the year 1981.

The discrepancies among industries are quite sharp. Four industries—chemicals, electrical equipment, machinery, and aircraft—account for 65 percent of all basic research in industry. Size of company is an important factor, and the cumulative effect of a few large firms is significant. The basic research activity of Bell Laboratories, IBM, and GE in 1985 can be estimated very roughly at about $425–$475 million total, based on personal discussions with senior representatives of those companies. Since NSF estimates that industry conducted $1,875 million for basic research that year with its own funds,[5] these three companies accounted for approximately 25 percent of all basic research conducted and funded by industry. Hence, this handful of firms conducts a large percentage of industrial basic research.

The statistics on basic research include both government-funded and company-funded activity. The practical reason is that the data is available only in that form. The philosophical reason is that government funding of basic research normally rests upon some existing competence of people and projects, so that relation of industry sectors with regard to basic research is not seriously altered by using this data.

The basic research activity of a large corporation is normally conducted in the central R&D organization rather than the technical organizations of the separate operating units. Divisional laboratories tend to work on shorter term projects geared to the product and process needs of the division for the near future. The corporate laboratory conducts longer term programs not tied to a particular product schedule.

The culture and environment that characterize a central laboratory differ somewhat from those of divisional facilities. While every group must be sensitive to the role of technical activity in the company's current business and future growth, the central laboratory is encouraged to develop a technical base for products not yet in the planning cycle of current businesses. The personnel are therefore removed deliberately from the constraints of definable needs—a critical requirement for basic research.

Other groups within the central R&D organization are responsible for applying successful basic research in order to transfer it to the divisional laboratories. A rule-of-thumb is that the sooner an operating division becomes involved in a new development, the more efficient the process of exploitation.

One other interesting characteristic of industrial research is the relative amount of supporting personnel and facilities necessary to conduct R&D in an industry. A simple index for such support is derived by calculating the total cost per professional person. In 1981, this averaged $107,000 for all industry, from a high of $140,000 for autos to a low of $70,000 for the paper industry.

Like other indexes, cost per professional person does not evaluate whether an industry is "good" or "bad," but simply describes a characteristic of that industry. However, it is meaningful to compare a company with the industry average. It is particularly meaningful as a means of tracking trends within a single company, which can provide research management with an early warning signal.

INDUSTRY AND THE TECHNICAL ENTERPRISE

The numbers alone tell something of the enormous range of technical activities pursued by private corporations. Industrial research itself is a major economic activity in its own right and hence an important objective for regional economic development.

While the magnitude of the activity is stunning, the truly significant features that are, in a sense, at the core of this book, are the trend and the role of these activities in our technical enterprise. Both are related to the level of industrial R&D, but their significance for the process of technical change is more fundamental than the size of the effort.

There is first the growth of industrial research relative to all technical activity. This has been rising steadily in all the major developed

Table 4-4. Percentage of National R&D Funding by Industry, Major Countries.

Country	1970	1983
Japan	59	64
West Germany	53	58
United States	38	50[a]
France	37	42
United Kingdom	42	41 (est.)

a. NSF data indicates that the U.S. figure for 1983 is closer to 51 percent.

countries except the United Kingdom over the past 15 years, as shown in Table 4-4.

Industry conducts more R&D than the figures shown due to government support, but that added effort is not as directly related to business plans and utilization of the research. Nevertheless, focusing only on the R&D funded by industry, that activity exceeds 50 percent of the national R&D in the three largest R&D performers within OECD. The OECD data also shows that industry-funded R&D had exceeded 50 percent of the national R&D activity for most of the medium-size countries in Organization for Economic Cooperation and Development by 1979. (See Table 4-5.)

Table 4-5. Percentage of National R&D Funded by Industry, Medium-Size Countries.

Country	
Switzerland	72.7
Belgium	65.6
Sweden	59.9
Italy	54.5

Thus, those technical activities intended to strengthen present businesses and, most significantly, to provide the technical foundation for new products, processes, and services constitute more than half the total R&D expenditures of the major industrialized countries of the OECD. The United Kingdom and France are just under 50 percent in industrial funding, and there are activities today to increase that in both countries.

When the actual dollar levels of these expenditures are considered, the dominance of industrial research stands out much more starkly, since it is heaviest in the countries with the largest national R&D expenditures. Considering the five largest spenders—the United States, Japan, West Germany, France, and England—the first three constitute approximately 84 percent of their combined R&D in 1983. Hence, in these five countries, industrially funded R&D accounts for roughly 53 percent, driven principally by the large U.S. position. For those concerned with the relative status of the United States, these figures may be of interest. The total national R&D expenditures of the United States in 1983 was approximately 10 percent more than the combined expenditures of the other four major spenders. The industrially funded R&D of the United States was

Table 4-6. Growth of Industrial R&D Funding ($ *millions*).

Year	Total U.S. R&D	Total Industry R&D	Industry as % of Total
1953	$ 5,124	$ 2,245	43%
1960	13,523	4,516	33
1965	20,044	6,546	32
1970	26,134	10,444	39
1975	35,213	15,820	44
1980	62,618	30,911	49
1984 (est.)	96,975	49,375	51

Source: Table 5 in "National Patterns of Science and Technology Resources, 1984," NSF 84-311, 1984.

approximately equal to the combined industrially funded R&D of the others.

More fundamental than changes in level of expenditure are changes in approach both in the United States and abroad.

The momentum of U.S. industrial research in the past 25 years has been an amazing phenomenon. It has demonstrated a consistency in growth in comparison with expenditures of the federal government that is seen in Table 4-6.

During the 1970s, total government R&D funding slowed, industry growth continued except for a minor dip in the mid-1970s. Industry funding exceeded government funding in 1980 and accounted for more than half of all R&D in 1981. In 1984 and 1985 federal funding increased due to defense R&D growth, so that the percentage contribution of industrial research may fluctuate. Nevertheless, the absolute amount of industrial research continues to grow strongly.

A special feature of this growth is the extent of basic research in industry. The relevant data is shown in Table 4-7. The numbers are not precisely comparable with the preceding table, since the data on basic research in industry includes some government funding. This does not change the several points of significance with respect to the growth shown:

- Basic research in industry has become an important part of the total basic research effort. While it is only a small proportion (a few percentage points) of industrial research, it is a significant

Table 4-7. Growth of Basic Research in Industry.

Year	Total U.S. Basic Research ($ millions)	Industry Basic Research ($ millions)	Industry as % of Total	Industry Basic Research as % of Industry R&D
1960	$ 1,197	$ 376	31.4	3.6
1965	2,555	592	23.2	4.2
1970	3,549	602	17.0	3.3
1975	4,608	730	15.8	3.0
1980	8,089	1,325	16.4	3.0
1984	11,850	2,300	19.4	3.2

Source: Tables 6 and 8 in "National Patterns of Science and Technology Resources, 1984," NSF 84-311, 1984.

contributor on an absolute level, and is roughly 40 percent of the basic research effort at universities.

- Industrial basic research grew steadily during the 1970s. While the level of industrial research effort increased in short-term developments during that period, lowering the percentage of basic research in total industrial R&D, the dollar level never dropped.

- The level of, and interest in, basic research by industry provides an important foundation for developing links with university research. The industry effort identifies areas of interest, and the industry personnel engaged in basic research provide an effective communication link between university research and industrial objectives.

Surveys taken by NSF in late 1984 indicate a sharp upturn in industrial basic research in microelectronics and biotechnology. Changing patterns in microelectronics result in expanding research within the companies, in research consortiums, in joint university-industry programs. In biotechnology, the major industries are expanding basic research as a foundation for conversion to biological processes where this is commercially feasible, and developing new products. Exxon, Monsanto, duPont, and others have followed initial cooperative efforts with universities by the establishment of substantial corporate laboratories.[6] The major pharmaceutical companies have expanded their own considerable research to build upon the new approaches inherent in biotechnical advances.

Hence, in those industries where future growth requires substantial technical advances or where wholly new opportunities are opened by technical advances, industrial R&D activity has expanded accordingly, including long-term commitments in basic research. The point is that this continuing increase in basic research is not simply an obligatory contribution to science or a mechanism for communication. It is a deliberate integration within the overall industrial R&D program of an activity necessary to growth in industries whose technical base is changing rapidly. The industrial activity in turn acts to stimulate that scientific growth outside the industry.

This indeed brings us full circle, reversing the conditions which created the need, the opportunity and the initiation of industrial research at the opening of the twentieth century. Industry had grown through invention and adaptation, drawing upon technical advances that emerged from a mixed reservoir of resources—universities, individuals, private laboratories, and government activity. The establishment of technical organizations within corporations focused the direction of R&D, increased available resources, and accelerated the process of technical change. The gradual integration of technical activity with corporate growth resulted in a massive expansion of industrial research. Industry now has the internal resources to drive the process of technical change, not merely to adapt and exploit what happens externally. But the nature of the technical enterprise is such that the pressure from any one sector affects all others. The momentum and resources of industry influence the directions of R&D at universities and the science policy of government. To the extent that industry is the agent in our form of society responsible for converting technical change into economic use, this is a natural consequence of technological evolution.

INDUSTRY–UNIVERSITY RELATIONS

This is a "hot" topic. In the past few years, it has been on the program of almost every professional technical society. There have been many major conferences on some aspect of the topic in most of the industrialized countries. It is the focus of much government attention, including specific funding from NSF for Industry University Cooperative (IUC) research.

Although interaction between universities and industry is not new, interest and support are greater than ever. One reason is obvious. The universities need greater and more stable resources from a mix of funding sources. Industry is willing to provide some support because it needs trained graduates and increased research. Another reason for the greater cooperation is broader economic pressures. Every country wishes to achieve economic development through the fullest use of all national technical resources, and to improve the international competitiveness of its products and its technical stature. Finally, a fundamental transition is occurring as the growth and momentum of industrial research creates new demands upon, and opportunities for, university research in parallel with the thrust of industrial activity.

Broader government interest throughout the world has been drawn to university–industry relations by the severe economic problems of the 1970s and the directions of economic activity in the 1980s. The direct pressure came from rising unemployment related to general recession and to declining markets for the products of heavy industry, the so-called smokestack industries. This was reflected in efforts to boost exports and a deep awareness that "high-technology" in-dustries—semiconductors, information technology, biotechnology—appeared to constitute the only islands of growth in a sea of declining markets.

Governments everywhere thus turned to technology as an option for economic growth. Policies and funding were focused on producing more technical activity, and accelerating the conversion of technical advances to economic use—that is, jobs. Since direct funding by government of R&D programs that could produce industrial growth is expensive and does not have an enviable track record, the one common denominator in all countries became the encouragement of university–industry interactions. This held out the hope for both strengthening the technical base related to economic growth and speeding up the transfer process from scientific advances to industrial developments.

These obvious factors may obscure a fundamental transition that occurred in the initiatives exerted throughout the technical community by industrial research and is changing industry's relations with universities. Industrial research in this century began as an effort to adapt externally accomplished technical advances for commercial purposes. It was not intended to create new knowledge, though it often had to do so. By definition, it was not self-sufficient technically. The subsequent growth, propelled by war, changed this (see

Chapter 6). From roughly 1960 to 1980 industrial research seemed self-sufficient technically with respect to its current businesses and its growth plans for several years ahead. It had sufficient technical knowledge and resources to support its growth plans. Even though the results of basic research are unpredictable, the process was purposeful and the probabilities of achieving applicable results were improved by careful selection and feedback. (Basic research is not a random process, though one may get that impression from occasional odes to serendipity.)

During this period of self-sufficiency, industry welcomed and encouraged relations with university research because the added inputs provided more options for progress, although it did not *need* the university research for ongoing business plans. The broader base of industrial research provided a greater number of opportunities for common research interests with universities, and this alone was a stimulus for such interactions.

Beginning approximately in the late 1970s and continuing today, a subtle shift appears to have occurred in industrial research. Many corporations are realizing they can no longer be self-sufficient technically. This is in part because the areas to be covered are increasing beyond the funds and personnel available, and in part because the efforts of any single corporation are becoming smaller in relation to the growth of R&D in all sectors and in all but the poorest developing countries. This is a natural consequence of the continuous growth in science and technology. Even the richest multinational corporation may not find it economical to pursue technical self-sufficiency. It can, however, identify the technical areas which *cannot* be pursued within the firm but which are relevant to the firm's strategic growth plans. They are very likely to be related to, or to feed into, some ongoing internal technical program, and certainly to potential growth plans which could not be implemented within the corporation itself.

These pressures have led to university–industry linkages involving far more substantial sums of money than were to be found even 10 years ago.[7] These include Exxon and MIT ($8 million over 10 years), Hoechst and Massachusetts General Hospital ($50 million over 10 years), Monsanto and Washington University ($23.5 million in 5 years), and many others.

While these large sums by individual corporations attract much attention, there has been a rise in activity in university–industry relations across the board. Many university research centers discussed in Chapter 2 have served to attract multiple corporate support for particular program areas such as robotics, biotechnology, and artificial intelligence. However long range and basic in nature, these programs serve to strengthen the current strategic growth plans of the companies involved.

The most striking fact about these interactions is that industry support for university research is overwhelmingly through the professional schools. Roughly 67 percent of this support goes to engineering disciplines, another 11 percent to medical schools, and 8 percent to agricultural schools. These three account for almost 86 percent of all industry support. Much of the support for biotechnology comes through the medical schools, and computer science is most often located in the engineering schools.

Another relevant point is that a great proportion of cooperative research programs between university and industry, roughly 70 percent, appears to have been preceded by some consulting arrangement between a faculty member and the corporation providing the support. This is not surprising, but it is an important factor in evaluating university policy with regard to faculty consulting.

The principal objective of industry in these relationships is to assure access to well-trained graduates. This is far and away the most frequently stated motivation. A less urgent objective is the specific research subject. Meeting this goal may mean establishing a specific research program at a university, but more frequently it means support for a particular area of research within a specific department or under the guidance of a reputable research professor.

An interesting corollary is that out of 214 cooperative university-industry programs studied, about 70 percent resulted from the initiative of the university. Industry was receptive, but more in terms of strengthening the field than soliciting specific research projects. There is no question about the increased interest of industry, but there is equally no question that the university is selling.

Finally, there is an interesting distribution of industry support for university research. In terms of dollar amounts, the top ten institutions in 1983 are shown in Table 4-8. On a national basis, industry funded $260 million of total university R&D of $7.26 billion, or about 4.5 percent. Thus, the top 10 institutions conduct $91 million, or approximately 28 percent of all industry-funded university R&D.

Clearly, industry support is attracted by competence and existing strengths, indicated by the size of total R&D expenditures at the re-

Table 4-8. Top 10 Universities/Colleges in Terms of Industry Funding, 1982.

R&D Fund Rank	University/College	Total R&D	Industry Funding	
			Amount	% of Total R&D
1	Massachusetts Institute of Technology	192,462	17,195	8.9
2	Georgia Institute of Technology	71,402	11,359	15.9
3	Pennsylvania State University	82,034	10,581	12.9
4	Stanford University	147,941	8,969	6.1
5	University of Arizona	88,247	8,554	9.7
6	Cornell University	145,769	7,495	5.1
7	University of Michigan	119,973	7,408	6.2
8	Purdue University	67,299	7,405	11.0
9	Virginia Polytechnic Institute and State University	39,909	6,515	16.3
10	University of Southern California	87,140	5,228	6.0

Source: Table B30 in "Academic Science/Engineering R&D Funds: Fiscal Year 1982," NSF 84-308, 1984.

search institutions. However, the percentage of university R&D funded by industry depends also on the emphasis placed by the university on developing these linkages. This is particularly true for those circumstances in which the university accepts a significant role in regional development. When we examine the top 200 academic research institutions, the 10 with the highest percentage of industry-funded R&D in 1983 are given in Table 4–9.

Obviously, these institutions do not correlate well with those conducting the most industry-funded R&D. The two tables should offer some sense of the sharply skewed distribution of industry funds to universities, and the wide differences in the impact of these funds at particular universities compared with the national average of 4.5 percent in 1983.

Thus, as discussed in Chapter 2 with regard to university research generally, there is a great variety in both the amount and the relative importance of industry funding for research among the wide mix of universities in the United States. When interested observers from other countries attempt to evaluate American activity in university–industry research interactions, should they focus on Yale or MIT or Georgia Tech, or perhaps Swarthmore or Wesleyan? The answer, of course, is that all must be considered, since the different tiers of educational institutions develop quite different relationships. Again, the *diverse* nature of research universities in the United States does not permit a simple averaging of numbers, since their separate educational and research philosophies provide healthy qualitative differences in their approach to industry.

This background of diversity is necessary to answer current questions about the growing intensity of university–industry relations. Do industry interactions affect the direction of university research? Will this activity distort the values and functions of the university?

The overall answer is that the university *system* remains healthy, independent, and in control of its own values, including the appropriate mix of education and research. Many institutions seek close involvement with industry, with 30 percent and more of their R&D funded by industry. Most research institutions welcome increased cooperation with industry, and will consider reasonable liberalization of patent rights, some publication delays, exchanges of personnel, and so on. Some of the largest and oldest research institutions have made clear that industry research funding is acceptable, but under

Table 4-9. Top 10 Universities/Colleges in Percentage of Industry-Funded R&D, 1982 ($ thousands).

R&D Fund Rank	University/College	Total R&D	Industry Funding	
			Amount	% of Total R&D
1	Worcester Polytechnic Institute	4,006	2,234	55.77
2	Desert Research Institute	5,798	2,196	37.88
3	Brigham Young University	4,375	1,328	30.35
4	The Medical College of Pennsylvania	5,867	1,644	28.02
5	Stevens Institute of Technology	4,691	1,290	27.50
6	Thomas Jefferson University	9,345	2,363	25.29
7	Michigan Tech University	8,772	1,673	19.07
8	Virginia Polytechnic Institute and State University	39,909	6,515	16.32
9	Georgia Institute of Technology	71,402	11,359	15.91
10	Colorado School of Mines	5,897	900	15.26

Source: Table B30 in "Academic Science/Engineering R&D Funds: Fiscal Year 1982," NSF 84-308, 1984.

traditional criteria that would not alter the complete control of the university over patents, research direction, publications, and the like.

There is, in other words, a market condition that sets the patterns for university–industry relations subject to general rules:

- Each educational institution sets its own goals concerning the balance between education and research, and this identifies acceptable conditions for research interactions with industry.

- No educational institution abdicates control over faculty appointments or independence in curriculum or direction of research.

- Industry has every incentive to maintain the strength of university basic research, and particularly its independence to identify and pursue the most fruitful research directions.

Though industry interactions affect the direction of university research, this is not the same thing as taking control of university research, or removing freedom of choice from the university. The increased breadth and magnitude of industrial research are transmitted to the university community through many forms: meetings, visits, publications, personnel exchanges. The direct support of industry for university research is another mechanism for exposing the university community to the technical needs of industry and to the current research pressures within industry. In certain fields, such as computer sciences, industrial research is pushing the frontiers of science and technology. An enormous attraction is exerted by such industrial research activity on the research plans of universities.

The more general statement is this: Industrial research affects the direction of university research in many ways. Industry funding of university research is a direct mechanism for attracting efforts to areas of particular interest to industry, even though such support is very often the result of proposals made by university researchers. But the underlying influence derives from the increasing breadth and vigor of industrial research, its identification of new areas for growth in science and technology, its leadership in many of these areas, and its opportunities for new graduates. All of these factors produce a combined pressure so that university researchers, acting through their own freedom of choice, are more attracted today to basic research that is of interest to industry.

What determines the choice of research project by a university researcher? There is first what is referred to as the "internal dynamics" of a field of research. A major new concept or new breakthrough may open the door to wholly new approaches to a wide range of subjects as, for example, the developments in relativity theory and the particle-wave character of electromagnetic radiation in the early part of the twentieth century. There is the convergence of knowledge and of instrumentation that makes experimental progress possible and provides a base for new theory, conditions that set the stage for the marked intensity of effort in biotechnology over the past decade. When these internal dynamics are favorable, then many research opportunities are open and reasonable progress is almost assured. There is much cream to be skimmed, and this serves to attract researchers.

A second factor is the interest of the outside world—the public sector and the private economy—in following through on particular areas of basic knowledge. This is an important factor to graduate students, and it is of considerable significance to the university researcher. There is the satisfaction in knowing that one's work is valuable to others, there is a sense of contribution to society, and there is the intellectual satisfaction from the technical exchanges that occur. Activity and interest in the world outside the university can be a powerful stimulant to basic research within the university.

Finally, there is the availability of funds. Given a choice of reasonably interesting challenges in different research areas, there are obvious advantages in selecting those for which financial support is available. Research facilities, support for graduate students, opportunity for travel are all valuable tools in the conduct of research. With them, research productivity can be enhanced.

It is with these factors in mind that I stated that industrial research activity generally, and industrial funding in particular, influences the direction of university research. Industry represents a source of attraction, a "pressure gradient" that makes it more likely for researchers to build up activity in preferred areas. This does not mean that a researcher devoted to a specific project is reluctantly "bought" by industry to drop that interest and take on something undesirable in order to earn a living. Anyone so devoted can normally find some support for the original project, though it may take more time and effort. Most university researchers are interested in areas of inquiry rather than specific projects; the same is true of most industrial interest in university research. Finding common interests is not too difficult.

For the most part, however, it is the external dynamics of a field, driven by the growth of industrial research, that exerts an overall influence on university research. The influence is largely implicit, and direct funding simply reinforces it. The increasing ability of industrial research to initiate technical change, including advances in basic research; to provide greater professional interactions with universities; and to offer attractive technical career opportunities for graduates are all major stimuli for university researchers.

It is certainly no coincidence that the sharp growth of the computer industry has been *followed* by increases in basic mathematics research at universities. This in no way minimizes the importance of the centuries of work in pure mathematics that provided the theoretical underpinnings of modern algebra and logic. Similarly, the steady and substantial research support of the National Institutes of Health in biology and biomedical research made possible the recent breakthroughs in biotechnology. But the commercial possibilities of these advances and the expanding research investments of those industries most immediately affected are resulting in an upsurge of university research in related areas. The interactions between the semiconductor industry and solid state physics provide still another example.

University research is influenced by the total environment in which research is conducted. Today a growing factor in this environment is industrial research.

The 70s: Universities Return to Industry

The story does not go back very far. Before World War II, university research activity was a faculty activity. External funds came from industry, private foundations, and philanthropists. These funds bought special equipment, permitted travel, supported some research assistants, and endowed buildings and chairs. The federal government was not a direct player.

In 1933 President Roosevelt made a request of Karl Compton, then president of MIT. The question was whether the federal government could and should take actions that would strengthen university research programs and capabilities. A study was initiated under the auspices of the National Academy of Sciences, through a board chaired by Compton.

This study emphasized the need for the United States to become more self-sufficient in basic science. The increasing national dependence on technology required a greater reservoir of basic science than could be imported from Europe. (It is interesting to compare our own dependence for basic science on Europe up to World War II with the situation in Japan vis-à-vis the United States after World War II.)

To develop our own self-sufficiency, the Science Advisory Board chaired by Compton proposed that grants in aid be given "to scientists already pursuing original research in the laboratories of academe and industry." There was no follow-up to this novel proposal in the early 1930s. Most relevant to our present situation is that the academic community of that day opposed the concept on the grounds that "Federal assistance on a large scale would pave the way to Federal control of their institutions."[8] The diverse grants from industry were desirable and unthreatening compared to the possibility of concentration of funding in a single sector. Presumably, nobody then could conceive of the magnitude to which such support would rise, but neither could they have pictured the multiple sources for such support within the federal government.

The experiences of World War II reversed these conditions for university research and the attitudes of university researchers forever. Universities performed brilliantly, effectively, and on schedule across the spectrum from basic research through engineering design and construction. From the glamor of radar and atomic energy to more traditional problems of ballistics and metal fatigue, the universities became an integral part of the national technical enterprise. The universities showed what their researchers could do under appropriate conditions and motivation.

This truly productive performance during World War II was conducted in close cooperation with agencies of the federal government, chiefly the military and the National Defense Research Council (NDRC), formed in 1940 to attract involvement of the nation's scientists. University research facilities and personnel expanded to such an extent that a sudden cut-off of the funding and relationships at the close of the war could have had disastrous effects on the university structure.

Two factors prevented this collapse and set the pattern for a new growth of university research. One was the still little-recognized actions of a farsighted group of individuals who persuaded the Navy Department to establish the Office of Naval Research (ONR) in 1946. While the charter spelled out the importance of basic research to the Navy mission, there is no question but what it was intended to provide some immediate support for university research until a slower transition to a different level of research could occur or new mechanisms could be devised. This action was followed shortly by establishment of the Air Force Office of Scientific Research (OSR) and the Army Office of Ordnance Research (OOR), now the Army Research Office. All were intended to support basic research, and all had the effect of providing continuing funds to the universities during the planning for, and the early growth of, the National Science Foundation.

A second factor was the expansion of relationships between the universities and the mission agencies of the federal government, making use of the experience with financial instruments and procedures developed during the war—grants, contracts, and the federally funded R&D Centers (FFRDCs), many of which were administered by universities. The Atomic Energy Commission became a major funding source for nuclear physics, while the Defense Department identified broad areas in materials, electronics, and mechanics for which financial support of external R&D was available. In time, other government agencies with a technical base developed similar relationships with universities, drawing upon many of the financial instruments developed by DoD and the AEC. The largest of these was the National Aeronautics and Space Administration, but many other agencies are active, such as the Bureau of Mines and the Department of Transportation.

Far from collapsing, as was feared by many in 1945, university research expanded upon these growing relationships with the federal government. The period of roughly 1950 to 1970 is sometimes referred to as the golden age of U.S. science, largely because of the growth of university research in those years.

All of this ferment had unintended, though hardly surprising, effects, however, on the direction, interests, and attitudes of university researchers, both faculty and graduate students. It strengthened those areas of basic research most relevant to the interests of the mission agencies. This thrust was intended and had many broad benefits. But the challenge of the scientific problems, the glamor of the national interests (atomic energy, space, missiles), the job opportunities, the availability of funds all combined to tilt universities toward the public sector and away from industry. Departments of metallurgy became departments or schools of materials science. Engineering schools became schools of applied science. What I have referred to earlier as the internal and external dynamics of the separate fields of research turned toward government.

Most important of all was the impact on graduate students and faculty, particularly the newer faculty coming into the universities in the 1950s and 1960s. From a purely practical view, funding was easier from government then industry. The effort required to obtain a $15,000 grant from industry might result in $75,000 to $100,000 from government. In fact, in those years industry did not often provide funds of $100,000 or more. If one wanted to think big, one thought of government.

Accompanying the growth of government support for university research was the increased activity in government research installations, in the federally funded R&D centers, and in the contractors carrying out the principal defense and aerospace programs. This meant positions in glamorous areas for new graduates. The more relevant the area of graduate research which they pursued, the better the position.

To conclude that all of this was a disincentive for university researchers to work with industry is an understatement. In all fairness, this did not produce a drifting apart. The traditionally research-oriented companies maintained liaison and hiring. A number of major institutions worked hard to continue industry relationships, including Columbia, Polytechnic University of New York (then Brooklyn Polytechnic Institute), Rensselaer Polytechnic Institute, Carnegie Institute of Technology (now Carnegie-Mellon University), and others. Many researchers in the engineering schools kept their industrial ties intact. However, the trends and statistics showed university basic research expanding in support of public sector objectives.

The factor most disturbing to industry during that period was probably the perceived attitudes of many faculty and graduate students about industry. While the late 1960s added a political tone which may have questioned the philosophy of private enterprise, a more serious concern was the apparent lack of understanding about the objectives and quality of industrial research. The belief seemed to be that if the government programs were glamorous, those of industry were mundane. While government research programs challenged the frontiers of basic science, industry needs were applied and incremental. The corollary, of course, was that first-rate scientists and engineers went to government-supported laboratories or stayed at universities (with government support), while those who had lesser qualifications went to industry.

Obviously not everyone thought this way, since industrial research in fact grew and strengthened in that period, and added many first-rate intellects. Nevertheless, there were enough indications of those attitudes, strengthened by the luxury of strong federal funding, that university–industry relations were at least harassed. Many meetings of professional societies in those years included sessions on the topic of strengthening graduate training for careers in industry. Invariably, the curriculum and research would get high praise, but industry representatives would express concern about attitudes. This rarely occurs today.

All of this changed just before 1970 and in the several years following. The steady growth of federal R&D funding slowed, and even declined for some agencies in constant dollars. While the absolute level of federal support did not go down, the annual increases of 15 percent and more came to an abrupt halt.

This is evident in the figures for the support of basic research by the federal government during the 20-year period from 1953 to 1973 shown in Table 4-10.

Table 4-10. Average Annual Rate of Growth of Federal Government's Basic Research Support (%).

	1953–61	1961–67	1967–73
Current dollars	17.3	17.1	3.0
Constant dollars	14.9	14.9	-1.1

Source: "National Patterns of R&D Resources: 1953–1973," NSF 73-303.

The impact on the plans and perceptions of university researchers and research administrators was drastic. Coupled to this was a cutback in defense contracts, which hit the Los Angeles area especially hard. Job opportunities for scientists and engineers in the aerospace and defense industries declined or disappeared in the early 1970s.

We know that all good things come to an end. Still, the changing conditions must have appeared to create an unusually cold, cruel world to the faculty who came into the university system during the golden age and to the graduate students emerging in the period from

1970 to 1974. They had entered a university research structure where funding at reasonable levels was easily available with only minor tailoring of research concepts to the broad needs of the public sector. Suddenly, these conditions were no longer valid. Government support was not automatic, and careers in government or contractor laboratories could not be relied upon. A painful uncertainty and uneasiness entered the university community. There was serious concern at the top level of university administrations, sensitive to their reliance on overhead income generated by government-funded research.

Not surprisingly, after 1970 university representatives began to make more frequent appearances at meetings which included the industrial research community or gatherings of industry representatives called for that purpose. Such meetings were characterized by a painful talking down to the audiences, a classic error in salesmanship. The gist of the university researchers appeals was that industry needed basic research, the university was the source, so industry should just provide the modest funds requested and leave the researchers alone. True, this approach was often dressed in various programs and mechanisms, but the bottom line did not vary. The impression conveyed by the university community was that the overall technical system was fine except for a shortage of funds for university research. Questions of relationships, priorities of research, functions of the university did not seem to be critical issues for the universities compared to the slowdown of government support.

This was highlighted by an incident in the early 1970s. Several actions were considered by the federal government to stimulate economic growth by drawing upon the nation's technical resources. Among other actions, legislation was advanced that would increase funds for university research presumably to compensate for the slowdown in defense research. The president's science adviser at that point was Lee DuBridge, a distinguished physicist and former president of the California Institute of Technology. DuBridge and his staff wanted a broad base of support for this proposed increase in university research funding. They apparently had the impression that the industrial research community was lukewarm, or at least not sufficiently outspoken, in behalf of the proposal. A small meeting was convened on the subject through the Industrial Research Institute (IRI). About a dozen senior technical executives of major U.S. corporations were there with two or three representatives of the White

House Office of Science and Technology (OST). The discussion went something like this:

OST: "Is industry in favor of a strong basic research activity at universities?"

Industry: "Absolutely."

OST: "Does industry favor increased funding for university basic research."

Industry: "Absolutely."

OST: "Then what is the problem?"

Industry: "During the past 20 years, university research has grown enormously, and distortions have entered the process—distortions in priorities, in procedures, in interests, and in career directions for graduates. If the question is "Will industry support added funds to continue these distortions?, the answer is no. If the question is broadened to whether industry favors greatly increased basic research at universities after careful balancing of all university objectives, with constructive attention to the scientific base of both public and private sectors, the answer is of course yes."

That conversation caused at least some careful thinking in the government and ultimately in universities. In any event, the relation between universities and industry steadily matured and strengthened during the 1970s. The proposals by universities became more realistic. They began to take into account the needs and procedures of industry. Mechanisms were proposed that included continuing research cooperation. Gradually the career patterns of graduate students shifted. In time the universities became more aware of, and sensitive to, the dynamic research growth that had been occurring in industry, and which now provided both stimulation and partnership.

Familiarity can improve relationships and understanding. But there are still generation gaps. As recently as the early 1980s, at a monthly luncheon meeting in New York of the Directors of Industrial Research (DIR), the talk was by a very famous and distinguished basic research scientist in the classic university tradition, who presented the case that industry's long-term future depended on continuing activity in basic research. Since the talent and sources of knowledge were in the universities, while industry assigned their technical people to short-term development, he said, then clearly the system would work best for everyone if industry provided funds to the universities to protect industry's future. In other words, with "your money and our brains" each party will be contributing its true strengths. The 30 or so diners at lunch were the senior technical officers of some of the largest and most research-intensive companies. They represented roughly 60 to 70 percent of U.S. industrial research and probably more of the basic research in industry. This group had presided over the growth of the major industrial research laboratories in the country, absorbing many of the best new graduates in science and engineering of the previous two decades, and advancing frontiers across the entire technical spectrum. The group was too experienced to be offended, but the naivete of the speaker in implying that that collection of research executives *depended* on the superior research talent of universities was a bit disturbing. In deference to the speaker, the point was not debated, though it caused much undercurrent of conversation privately. One representative of a major oil company did feel moved to ask the speaker if he recognized that some very bright people could be found in industrial laboratories and some very good research went on there.

The preceding incident notwithstanding, a healthy and constructive relationship had developed by the 1980s between universities and industry. Government still plays the major funding role, but industry R&D activity reinforced by industry funds has been important in achieving balance in university research and developing new insights among faculty and students about industry.

Strengthening the Bonds

What trends will affect the future of university-industry relations and, if they serve society well, what can be done to strengthen them? Will the cooperation and linkages simply increase, or will new forms arise?

The universities came to industry in the early 1970s seeking funds. By the 1980s they were developing partnerships. Perhaps the most important factor in the evolving nature of these interactions is the willingness of each party to understand and accept the objectives and values of the other. The quality and vigor of industrial research activity provide a strong base for intellectual exchanges that had not been perceived or appreciated by universities in earlier years. There is now a strong foundation for growth.

This is a new ballgame for industry as well. As new working relationships evolve, industry has the challenge and the opportunity to develop its technical strategy so that university linkages become an integral part of the total technical resources drawn upon for future growth while preserving the objectivity and freedom of choice essential to the functions of a university. Since this was achieved when government funding rose from zero to almost 70 percent, it should certainly be possible to maintain those features even if industry funding doubles (from roughly 4 percent to perhaps 8 percent).

The increased exposure of university faculty and students to the programs and researchers within industry serves as a stimulant to the university research community. But the industrial programs tend to advance along interdisciplinary lines. This does not mesh neatly with the departmental structure of universities. Hence, there is a steady pressure for universities to create mission-oriented centers. The intent is to permit broad approaches to a field, such as robotics. The effect will be to facilitate cooperative research with industry.

This is a constructive trend for reasons that were discussed in Chapter 2. The need and the opportunities have been recognized by the National Science Foundation, which has supported a number of mission-oriented research centers, and has just launched a new program of Engineering Research Centers. All of these anticipate industry involvement and funding.

Industry will expand research support at universities with traditional grants to individual faculty in departments and with program funding of the research centers. Each university will decide separately whether it wishes to build substantially in a particular area that can attract industry support. If so, more mission-oriented centers will be launched or expanded.

Such centers will raise policy issues for the universities. If the growth of a center requires some commitment to long-term stability, then at least some senior personnel should have tenured appointments tied to the center. This is no problem when the center is predominantly within a single department, but it conflicts with the departmental structure for truly interdisciplinary subjects.

Institutions which commit themselves to long-term industry partnerships will very likely make such decisions. This in turn places an obligation on the relevant industries to accept long-term financial commitments. Not all universities will move toward this two-way street. Our large and diverse university system is an insurance that no single direction will be followed by all.

An additional benefit arises from the technical initiatives of industry. As recently as perhaps 20 years ago, the university system was the base for the diffusion of knowledge throughout the scientific world. Thus, the level of knowledge was kept fairly uniform in all developed countries, although the capability for generating or using the knowledge varied. Today, industry is creating technical links throughout the world through subsidiaries, joint ventures, international consortiums, and extensive licensing. This is producing a worldwide flow of technology, analagous to the earlier worldwide flow of science.

The university-industry relationships facilitate the exposure of universities to these technological advances, just as industry is exposed to basic scientific advances through the university network. This strengthens university research. It also, however, raises the potential for "haves" and "have-nots" in the university community based on the extent of industry contacts and cooperation. This is quite analogous to earlier observations that universities with extensive defense research contracts had better access to technical advances, resulting in greater excellence in research capabilities and initiatives in fields related to those contracts.

NOTES TO CHAPTER 4

1. George Wise, "Science at General Electric," *Physics Today* (December 1984).
2. National Science Foundation, "Science and Technology Resources," NSF 84-311, 1984, page 83, table 77.
3. *Business Week*, February 18, 1985, p. 84.
4. See W. Leontief, "Input–Output Analysis," in *Encyclopedia of Materials Science and Engineering*, edited by N. Bever (Oxford, England: Pergamon Press, 1986).
5. Table 2 in the NSF's report titled "National Patterns of Science and Technology Resources, 1984," 1985.
6. "The Behemoths Move into Biotechnology," *Business Week*, November 5, 1984, p. 137.
7. Lois S. Peters and Herbert Fusfeld, "Current U.S. University Industry Research Connections," National Science Foundation Report NSB 82-2, 1982.
8. Milton Lomask, "A Minor Miracle: An Informal History of the National Science Foundation," NSF 76-18, 1976.

5 GROWING COMPLEXITY OF THE TECHNICAL ENTERPRISE

The massive technical activity now in progress throughout the world represents just one more stage in the evolution of science and technology over thousands of years. Perhaps less familiar is the notion that, as science and technology advance, each successive step becomes more difficult. Thus, continuing technical progress calls for increasing resources—more money and more personnel. Assigning these resources to the increasing numbers of different areas of science and technology in turn calls for more organization.

The concept that continuing technical advance is progressively more difficult is not obvious at all. Nevertheless, we must grasp precisely how technical progress depends on increasing resources in order to understand the role of industrial research and its impact on the growth strategies of industry.

It is fairly easy to find examples of the obvious. As we try to see smaller objects, we have gone from optical microscopes to electron microscopes. To see greater distances, we have advanced from optical telescopes to radio telescopes. Routine laboratory instruments have changed in the last 75 years as voltage meters were replaced by oscilloscopes, which were superseded by computers. Knowledge of atomic and nuclear structure emerged first from chemical reactions and their implications for the periodic table of elements, advanced to the use of natural radioactive materials, to x-ray generators, and then to the present massive particle accelerators that require square miles of land area and the cooperative activities of nations. These trends call for more people, more expensive equipment, more organization.

Far more uncertainty is attached to the not-so-obvious features of the increasing difficulty inherent in technical advances. How can we compare different technical advances? Can we define a "unit" of technical advance in order to evaluate trends in the cost per unit? How do we compare the significance or value of different technical advances?

All of this may seem to be the substance for a seminar in philosophy. These questions can, however, be approached quantitatively. In the process of doing so, we can develop a theoretical basis for the structure and magnitude of our overall technical enterprise.

There is an absorbing book which does indeed address boldly the philosophical issues and quantitative analysis to which they lead. By Nicholas Rescher, it is titled *Scientific Progress: A Philosophical Essay on the Economics of Research in Natural Science.*[1] The text is more esoteric and mathematically detailed than a nonspecialist may desire, but it has broader implications for industrial growth than might be considered by the historians of science who are the expected readers, and possibly more than was intended by the author. Rescher has made a major contribution by integrating the concepts of many thinkers to present a coherent approach to the issues, and much of the discussion in this section and the next is drawn from his efforts.

Rescher describes a number of attempts to measure growth or complexity of the technical system. The trend toward multiple authorship of scientific papers is one measure. A study in the field of chemistry shows that during the twentieth century, the number of authors of papers on chemistry has been increasing by 10 percent every 10 years. Papers by a single author now account for only 17 percent of chemical literature, and will be almost zero by the year 2000.

An interesting and valuable relationship of greater generality is the square root rule for "elite" populations within a general population of people or objects or concepts. Known as Rousseau's law, this relation states that the elite members of a group amount to the square root of the numbers of the group.[2] For example, taking the number of colleges and universities in the United States as approximately 2000, the number of major universities is on the order of 45, the square root of 2000.

The general form of this relation is that important results are proportional to the square root of the number of results. The specific relevance to R&D is the distinction between scientific progress (the measure of significance) and scientific output (the number of events).

The rising cost of research is indicated by some general calculations. An estimate by D. S. Bromley is that the intrinsic cost of carrying out scientific research increases by 3.5 percent to 7.5 percent per year apart from inflation.[3] A survey of 17 laboratories showed that the cost per professional researcher doubled from 1950 to 1960, a period of relatively low inflation. While total fixed dollar costs multiplied by 4.5 from 1950 to 1960, R&D output only doubled, so that the fixed cost per project roughly doubled.[4]

A fundamental factor in the increasing complexity of R&D is the continual opening of new fields of knowledge. An impressive example is Rescher's model of the evolution of physics in the past 60 years (based on growth rate of findings) given in Table 5-1.

Table 5-1. Taxonomic Evolution of Physics, 1910-1970.

Taxonomic Level	Number of Units at This Level				Growth Rate	
	1910	1930	1950	1970	Per Year	Per Decade
Physics	1	1	1	1	Linear Growth	
Branches	9	11	14	17		
Specialties	20	43	95	210	4	48
Subspecialties (problem areas)	60	140	350	850	4.5	55
Problems	300	800	2,100	5,600	5	63

Source: N. Rescher, "Scientific Progress," University of Pittsburgh Press, 1978.

These and many other examples drive home a fundamental feature of the conduct of science and technology. There is a steady increase in breadth, in costs, in R&D interaction. In all this, important results grow much more slowly than overall output.

WHY THE GREATER COMPLEXITY?

The numbers that describe the patterns of growth are intriguing. Some may even be useful for planning and forecasting. Nevertheless, it takes more than description to satisfy our intellectual curiosity. We want some form of cause-and-effect explanation, even some plausible hypotheses as to why these patterns exist. Then we can anticipate how the increasing complexity inherent in technical change can affect the technical system itself. There is even the possibility that improved understanding of this complexity may suggest modifications of the technical system that can either minimize the anticipated complexity or come to terms with it. This has considerable implications for the costs of R&D, the allocation of people, and the structure of technical organizations.

A technical problem or objective that is considered to be "big" or "complex" or "difficult" falls into two categories or a mixture of both, called by Rescher "complexity-intensive" and "power-intensive." He defines the two categories as follows:

- *Complexity-intensive problems* can be broken down into components to be pursued by separate groups of researchers. This is the essence of systems development pursued in large military programs, and it is typical of interdisciplinary research. This category is associated with the large, diversified laboratory.

- *Power-intensive problems* cannot be approached in a series or combination of component projects, but require a substantial minimum concentration of resources to make any progress. This would apply to a particle accelerator, nuclear reactor, or the powerful rockets and launching systems necessary to place a satellite in orbit.

The overall space program is an example of both categories. It is complexity-intensive in that a great many disciplines are called on for necessary contributions in separate projects—electronics, materials, astronomy, life-support systems, and so on. Nevertheless, the ability to launch a rocket requires a power-intensive component.

Obviously, this is a matter of degree. As a large program is divided into components, each calls for an allocation of resources. If a component problem cannot be subdivided, then the question is whether the resources necessary to resolve it are appreciably greater than what would normally be available in the conduct of R&D. If so, then that component defines the power requirements of the overall program.

As science and technology advance, individual research groups more frequently encounter problems that require more complexity or more power than its resources can provide. This is obviously true for an individual researcher or a small group of researchers. It holds for research centers, and it applies equally (though at different levels) for the largest industrial research laboratories or national laboratories. There are several consequences.

1. As each branch of study requires data calling for more sophisticated measurement techniques, there is increased cost associated with that branch and there may be increased pressure to enlarge the research organization. It may also lead to greater specialization; that is, the number of groups pursuing that field may decrease. In special instances, it can lead to cooperative efforts, as in the establishment near Geneva of CERN, the European Nuclear Research Center.

2. As a branch of study broadens, a particular research group or laboratory can pursue only a small portion of the field at a given time. Any complex program can be pursued by breaking up the problem into pieces and applying a research group's limited resources to the pieces a step at a time, but this stretches out the time necessary for completion. When time is critical, the organization must increase the resources available to it or combine with other organizations in order to assemble sufficient resources.

3. Since technical advances in areas of increasing complexity often require breaking down the process into a number of simpler components, the probability that any one research project will be highly significant is reduced. In other words, breadth and complexity lead to an increasing number of projects. Significant advances then occur at a much slower rate, since each advance requires completion of two or more projects whose results are interdependent.

Underlying the statistics on increasing cost, people, and organization required for the continued production of significant advances is the relation that scientific effort increases at a greater rate than high-level scientific progress.

The rising costs derive in part from the relation of advances in knowledge to the technology needs of problems requiring concentrated resources, and in part to the increased time and personnel required by the division of effort in complex areas. The growth in people and organization derive from the efforts needed (1) to create

Table 5-2. Patent Output in the United States, 1900-1954.

Year	Scientists and Engineers	Patents Granted	Patents per 1,000 Scientists and Engineers
1900	42,000	24,660	587
1910	86,000	35,168	409
1920	135,000	37,164	275
1930	227,000	45,243	199
1940	310,000	42,333	137
1950	573,000	43,072	75
1954	691,000	33,872	49

Source: Fritz Machlup, *The Production and Distribution of Knowledge in the United States* (Princeton, N.J.: Princeton University Press, 1962).

the increasingly resource-driven technological developments, (2) to combine technological and scientific efforts, and (3) to tie together those separate projects that make up a significant advance.

The sense of a broadening technical base and of escalating costs is summed up by the great physicist Max Planck, formulator of the Quantum Theory, quoted by Rescher: "With every advance in science the difficulty of the task is increased; ever larger demands are made on the achievements of researchers, and the need for a suitable division of labor becomes more pressing."

Much of the preceding discussion refers to "advances" and "significant findings." Both terms are subjective. Attempts to be quantitative are possible but the results are difficult to interpret. For example, data gathered by Machlup on patent productivity shows a steady decline in output per professional technical person. (See Table 5-2).

The trend is clear though the data does not account for lag time (years between the research and the patent), and it does not identify the number of professional people in R&D specifically. The last few numbers certainly have something to do with the sharp increase in R&D after the war and less time to file for patents during it. However, this type of data combined with subjective judgments give some support for the concept that "advances" and "significant findings" do not increase proportionately with increases in resources.

Rescher draws upon a number of analogies in different fields (particle acceleration, the Weber-Fechner relation in psychology, and the Richter scale for earthquakes) to propose a logarithmic

Findings – Distribution Law. If a range of effort is given by E (dollars, personnel, number of projects), and the number of first-rate findings is given by F, then F increases proportional to the logarithm of E. That is, $F = k \log E$, where k is a constant that depends on the variable chosen for E. For those whose use of calculus may be rusty (including the writer), this relation means that the effort E must increase exponentially with time in order to keep the rate of first-rate findings, dF/dt, constant. E must increase from 10 to 100 in order to multiply F by 2. Putting this the other way around, if we want to produce 3 times the number of significant findings in a given time, we must increase effort 1,000 times. These numbers spell out dramatically the significance for the overall technical system of the continuing process of technical change.

To close this section, let us return to one aspect of cost escalation: the relation between science and technology. There is a twofold dependence of science on technology. First, increasingly sophisticated equipment and technologies are necessary to obtain data that can guide scientific theory. Second, the ability to obtain data beyond past limitations provides insight that can lead to wholly new concepts. In addition, new instruments can improve the productivity of research and thereby accelerate the entire process, as we see in the widespread use of modern computers. J.B. Conant, distinguished former president of Harvard University, remarked, "Science owes more to the steam engine than the steam engine owes to science."[5]

Thus, advances in knowledge depend upon new instruments and tools that rise steadily in cost as the performance characteristics become more severe. The fact that the cost of computation has dropped sharply so that automated analyses have increased productivity in some research fields contributes to slowing the accelerating costs of research.

Consider some recent additions to the technical arsenal, and the effect of performance on cost. The familiar optical telescope with a 10-inch diameter mirror costs about $10,000, but this rises quickly to about $10 million for one with a 300-inch mirror and greater resolving power.[6] The general phenomenon of pushing the capabilities of complex equipment to continually greater limits is seen in the evolution over time of nuclear magnetic resonance equipment. Rescher shows that in the period from 1954 to 1970, the cost of this equipment went from about $20,000 to roughly $300,000 in constant dollars due to continuing advances in its characteristics.[7]

Finally, we should take note of very similar cost behavior in any process that attempts to achieve some theoretical limit, a situation often encountered in risk analysis as we try to reduce risk to zero. Rescher depicts the cost escalation encountered in the process of approaching zero pollution. The effect of reducing demand for biological oxygen in decomposing organic waste from a beet sugar plant.[8] As the demand is cut by 30 percent, then 95 percent, the cost per pound of waste removed changes from less than $1 to about $60. Similar curves exist in all areas of pollution reduction, as in removal of sulfur dioxide from smelter emission or fossil-fuel plants for electric power generation.

In summary, then, advances in science or technology or in their applications require constantly increasing resources, either to maintain the historical rate of technical change or to maintain desired advances in the quality of life. These characteristics lead to policy issues for society as a whole. How much of a nation's total resources should be devoted to continuing or increasing a given rate of technical change? Where will the resources come from? What form of organization can accommodate them?

COPING WITH COMPLEXITY

Two things seem clear about modern capabilities to advance science and technology. First, the rate of technical change is surely at least as high today as in any period of civilization to date, so that we have in fact been able to increase resources sufficiently to maintain and even increase that rate. Second, the organizations and procedures which are used in the conduct of R&D are greatly different today than a century ago, and the conduct of R&D throughout the modern industrial age is different than it had been in the evolution of science and technology prior to the Industrial Revolution.

The state of science and technology in the industrialized countries in the nineteenth century was such that continuing advances required substantial concentration of people and facilities to attack many of the identifiable problems and opportunities in mechanics, chemistry, and electricity. The demands for concentration of resources in the different fields of science and technology were increasing.

Thus, in more and more areas, progress required organized laboratories. The primary continuing research efforts were at universities, and the aggregate programs of faculty research with the associated

support structure (machine shops, glass-blowing facilities) constituted much of the resource base necessary to maintain the rate of
technical change. The growth of university research proved adequate
for the continued advance of science, and for developing much of the
base of the technologies required for the chemical and electrical
industries.

As industry absorbed these advances, the technical horizons which
opened could not be supplied by university research, and indeed that
would not have been appropriate. The growth of organized industrial
research at the start of the twentieth century was therefore one of
those events which, if it had not occurred, would have had to be
invented.

Some new mechanisms for assembling, concentrating, and funding
technical resources had to be developed in order to maintain the rate
of technical change required by the expansion of technology-based
industry.

To decide how much of its resources to devote to R&D, a firm
may try to estimate the future income that might be generated by
the results of R&D. But two other features of industrial research
were absolutely essential to provide substantial and growing technical
resources that would at least sustain the rate of technical change
present at the beginning of the twentieth century.

First, industrial research has always been an integral part of an
income-producing system. The ability of a corporation to increase
the probability of return from R&D by careful selection of research
programs, and the development of a patent and legal system that can
assure reasonable appropriability of the R&D results for the corporation, justify and make possible the very considerable funds necessary to operate a large research laboratory today. Connecting the
R&D effort to the source of funds generated in part by the results of
R&D was a critical mechanism to answer most of the public policy
issues raised by the increasing level of R&D.

Second, the increase in technical resources is a tangible physical
process, not just a phrase. It means the concentration and coordination of research personnel, supporting staff, equipment, and the
handling of large sums of money. This calls for organization or, more
explicitly in this subject, research management. The framework and
the culture for this effort were established easily within industry.

There is, of course, another major source that can provide the necessary growth in technical resources needed to maintain a desired
rate of technical change—the federal government—and it has done

precisely that in particular circumstances. The government has been a principal factor in the supply of trained scientists and engineers. It has concentrated technical resources whenever problems in the public sector called for an increased technical base, problems related to defense, space, or energy. And in research areas of high "power-intensity," it has most often been the federal government that provided the concentration of resources needed for progress, for example, particle accelerators.

The continuing increase in technical resources led to greater concentrations of effort to pursue fields of increasing difficulty. These concentrations exceeded the limits of independent inventors and researchers, and also exceeded the limited resources of universities. Thus, whether completely fortuitous or resulting from indirect pressures, the growth of industrial research provided a mechanism that maintained the historical increase in technical resources.

A mathematical law does not control events; it merely describes them. When the observation or hypothesis is advanced that the number of important results is roughly the square root of the total number of results, or that the number of important findings is proportional to the logarithm of the resources allocated, the mathematics describes the results of actions. It is not a determining cause.

To be more specific, industrial research laboratories did not emerge and grow because anyone examined the number of major scientific advances in the preceding decade and ordained a doubling or tripling of technical effort in order to achieve the same number in the next decade. At any point in time, the breadth of science and technology stimulates ideas for application, for invention, for new directions of research. The growth of industry provided new justification for a great many concentrations of technical effort. These in turn led to more opportunities, more attention to the scientific foundations, more ideas for application, and so on. The level of science and technology by 1900 identified technical opportunities which required concentration of effort, organized industrial research provided a new mechanism, and this has raised science and technology to new levels. In a broad statistical sense, this process is described mathematically by the mathematical laws discussed.

Every mechanism that requires exponential growth must saturate in a finite world. Nothing requires a mathematical hypothesis to hold for all situations. At some moment either we must advance another hypothesis to describe the results of saturation, or we must evolve new mechanisms of growth. Some new mechanisms are suggested in the following section.

DIRECTIONS FOR THE TECHNICAL ENTERPRISE

The organizational structures which have emerged for the conduct of R&D serve to optimize technical progress. The growth of science and technology is a growth in numbers of people, numbers of projects, numbers of new branches and subbranches of technical disciplines, and numbers of locations. The steady, though slower, advance in the number of truly significant findings occurs in three different ways.

1. By concentration of resources necessary to attack "power-intensive" problems

2. By random chance, given that out of every n number of projects, roughly $\sqrt{n}$ are likely to have a greater significance than the rest

3. By the combination of results from a number of separate research activities which provide the concepts or data that permit the cumulative perception leading to a new process, property of material, invention, physical law, or mathematical theory

Implicit in all of these mechanisms in order for progress to occur is the requirement that communication take place. There must be communication of results among related activities so that the separate projects, tasks, and personnel can reincorce one another and have their outputs coalesce into a single theory or application. Thinking about the physical actions necessary to accomplish this, we begin to develop a sense of the escalation in cost, complexity, and organization as the base and magnitude of science and technology have grown.

Progress in science and technology is like a jigsaw puzzle, with hundreds of separate tasks or projects needed to advance a particular field.

The real world of R&D is far more complex. It is a dynamic system changing with time. The 100 or 1,000 separate projects are progressing at different rates, and the results emerge in varying amounts over time. Further, the different activities are dispersed geographically, and many of the relevant projects may not be known at all to other researchers who might benefit from the results.

The early evolution of science and technology may have progressed substantially whenever the results of a few separate lines of research were brought together, but communication among these activities was infrequent and often difficult. Not everyone could make the effort of a Marco Polo. Today, we have an amazingly widespread and rapid system of worldwide communication. However, substantial progress at our more sophisticated level of science and technology requires contributions from such a wide range of technical efforts that we must place steadily increasing emphasis on mechanisms for communication and for analysis.

An important characteristic of the modern system of organized research, and one of its principal functions, is to convert this process from a random to a purposeful activity. Perhaps a more realistic statement is that the structure of a modern R&D laboratory and the overall structure of the technical enterprise raise the probability that related results from many sources and from different technical disciplines are brought together at a common time and place. There are three features which contribute to increasing the probability for such interactions, hence for achieving significant technical progress.

1. The interdisciplinary nature of large R&D laboratories, particularly in government and industry, brings together technical personnel from many disciplines, through projects or organization or both. Each researcher may represent the state of the art in a specialty.

2. Communication of any large R&D organization with general technical activity in the external world occurs importantly through the informal activities of "gatekeepers." The word was coined by Thomas Allen at MIT, who traced the flow of information into any research organization through those few individuals who developed extensive linkages with outside activities.[9]

3. Development and use of sophisticated computer networks provide access to multiple abstract services and data banks. For all practical purposes, a very large percentage of the world's published technical results is available through these networks.

Thus, the operation of the principal R&D organizations provides for effective access to relevant published materials in the ongoing conduct of R&D. This leaves two categories of linkages to consider with regard to future directions of the technical enterprise: access to unpublished material, primarily technological know-how, and co-ordination of R&D capabilities necessary to pursue future programs.

Technology, or technological know-how, is transferred through a great many channels, and we have seen marked increases in all of them. The most common, of course, is patent licensing, but there is also the device of technical exchange agreements between companies. The embodiment of new technology in a product or system which is sold constitutes an important form of technology transfer.

One example of the extent to which technology is exchanged through such commercial channels is to examine the sales and purchases of technology by each major country. Data compiled by the Deutsche Bundesbank gives the volume of this trade, shown in Table 5-3.

The table gives a sense of the flow and increase in technology transfers. The specific country data is not as important to this chapter as the extent of exchanges, though it shows clearly the strong imports of technology by Japan, West Germany and Italy, and the strong exports of technology by the United Kingdom and the United States. The main point, however, is that purchases of technology are substantial and spread throughout the industrialized countries.

We are accustomed to the free flow of science through traditional professional channels, which has been critical to the progress of science. There is, for all practical purposes, a comparable flow of technology throughout the technical enterprise, not free but available and accessible. Through the multiple channels of licenses, products, scientific instruments, and the like, technology flows throughout the world with minimum delay. Because of the interdependency of science and technology, this is especially critical to the progress of science *and* technology.

The growth and effectiveness of the overall technical enterprise depends not only on developing access to the existing reservoir of science and technology, but on bringing together the relevant competencies to approach new problems. The linkages to existing knowledge provides concepts and solutions for current R&D, but linkages to the necessary combination of technical capabilities can provide the resources for future R&D.

This is the final step and present challenge to the technical enterprise. The development of interdisciplinary, or simply multidisci-

Table 5-3. Receipts and Expenditures on Patents, Inventions, Processes, Copyrights, and Related Items (*DM millions*).

Country	1972			1980		
	Receipts	*Payments*	*Balance*	*Receipts*	*Payments*	*Balance*
France	277	904	-627	902	1,866	-964
Federal Republic of Germany	674	1,574	-900	1,101	2,624	-1,523
Italy	291	1,295	-1,004	1,565	2,365	-800
Japan	223	1,741	-1,518	643	2,411	-1,768
Netherlands	329	490	-161	760	1,166	-406
Sweden	66	199	-133	168	400	-232
United Kingdom	1,079	978	+101	2,185	1,681	+504
United States	8,833	938	+7,895	12,698	1,375	+11,323

Source: R. Rothwell and W. Zegveld, *Reindustrialization and Technology* (White Plains, N.Y.: Longman, 1985).

plinary, technical competencies within a single technical organization has provided an important factor in the continuing scientific and technological progress of the past hundred years. The ability to link such necessary competencies outside the individual technical organization may prove equally important to continuing that progress in the future.

There are many signs that this process is underway. Patent licensing is perhaps the most elementary form of such reaching out. Joint ventures and private consortiums, such as for development of a new jet engine or for ocean mining, are more active examples. The recent growth of university-industry research cooperation is another form of enlarging technical capacity. Within the past five years, several collective industrial R&D groups have emerged to pursue technical programs that will strengthen the competitive position of the member companies. All of these are tools for continuing technical progress.

WHAT IT ALL MEANS

In the internal dynamics of technical progress two lines of development emerge. One is the constant pressure to increase technical resources in order to continue the rate of significant technical change. The other is the emergence of organizational structures to coordinate and manage these resources effectively.

The organizational structures of the individual components and of the total technical enterprise not only will serve to coordinate the network of science and technology so as to facilitate technical progress, but will initiate actions that can increase technical activity without exceeding the limits of available resources. This will not occur by any deliberate plan, but as the reaction to the continuous pressure of current needs and the continuing growth of technical activity worldwide in industry, university, and government.

The mathematical description of technical progress stated that, when the technical effort increased exponentially with time, the rate at which significant technical findings emerged was constant. The physical actions that produce this relation are twofold: (1) conducting and merging results from increasing numbers of technical projects needed to produce significant progress due to the broadening of technical disciplines, and (2) concentrating substantial resources to overcome complex or difficult technical obstacles.

Taking the description of technical progress literally, we would require exponential increases in personnel and funds to continue our rate of technical advance. This is physically impossible at some point in a finite world, and it is very likely to be in conflict with other objectives of society long before resources are exhausted. Fortunately, it is probably not necessary to pursue such increases literally.

Technical progress can be maintained by actions which have the *effect* of increasing resources. The growth of organized R&D and the emergence of a broad technical enterprise makes it possible to have the impact of greatly increased technical resources with far less allocation of effort than would be forecast by simple extrapolation of growth curves.

This is not a matter of contradicting the mathematical description of scientific and technological progress as it has occurred prior to the twentieth century. Rather, it is a question of changing certain operating features that characterize the system for producing technical change, which can be accomplished by deliberate actions of the technical enterprise. To be specific, there are three factors in the process of technical change that can be influenced by such deliberate action.

1. *Shortening the time for merging results.* The identification of related technical activities worldwide and the instantaneous transfer of data through computer networks permit the interactions of science and technology, of theory and experiment, to proceed more rapidly today than 10 or 20 years ago. This increase in the rate of diffusion and of assimilation of technical advances has an effect similar to increasing the rate of technical effort with regard to the generation of significant technical findings within a given time.

2. *Decreasing randomness in worldwide technical activity.* The mathematical relation between significant technical progress and amount of technical resources used is a statement of probability. While the growth of research effort is not truly random, the circumstances under which projects are initiated and conducted in tens of thousands of R&D groups worldwide provide an independence of action that leads to the general probability distribution observed. The actions of large R&D organizations and of the total technical enterprise to identify broad opportunities in particular research areas, establish effective communication among active groups, and

concentrate funds for those research areas all act to decrease the randomness among separate efforts and increase the probability of significant progress for a given technical effort. This occurred for materials research in the 1960s, biomedical research in the 1970s and 1980s, and is happening in microelectronics today.

3. *Linking resources to permit broader approach to new areas.* The traditional growth process for a single R&D organization would involve continuing addition of people to provide contributions from newer subbranches of science and technology and from broader technical fields that bear upon new programs. It would also include rising investment in new equipment. Limitations of funds and people will ultimately slow this growth, so that the organization must focus its resources on narrower goals, or rely more on external results. The first path means lost opportunities, the second path means lost time. Initiatives can be taken, have been taken, and will be taken by R&D organizations to develop mechanisms for combining technical resources needed to pursue particular research areas. These may be long-term arrangements (merger or cooperative association) or relatively short-term (consortium or joint venture). It can involve a broad technical mission or a specific project. It can be an agreement on paper to use existing facilities or to build a new facility. The important point is that linking available resources to provide a broader technical base can have the effect of expanding beyond the financial and personnel limits of a single organization, while focusing the efforts of multiple sources within a common time frame.

The technical enterprise acts today to change all three factors. More links are created, more clusters of purposeful and relevant R&D activity are identified and encouraged, and more rapid communication and analysis of technical advances on a worldwide basis are being pursued effectively. We are therefore able to do more and learn more within a given period of time than we could through the steady expansion of existing R&D organizations alone, or the continued nucleation of new R&D groups. These growth mechanisms proceed, of course, but significant technical progress no longer depends on them alone.

The theory and mathematical descriptions of scientific and technical progress have derived from observations of traditional growth through expanding the reservoir of physical resources devoted to continuing R&D. I have suggested that comparable progress within a given time frame can be made through the creation of more linkages and increased frequency of communication.

To pursue these alternate actions, as we are indeed doing, requires organized R&D, even at universities. In the case of industrial research, combining technical resources calls for coordination of technical planning with business strategy at a high level. The pressures for doing so are increasing with the growth of the technical enterprise.

NOTES TO CHAPTER 5

1. University of Pittsburgh Press, 1978.
2. This relationship is discussed in the *Social Contract* of Jean Jacques Rousseau (1762).
3. D.A. Bromley et al., "Physics in Perspective," NRC/NAS, Washington, D.C., 1973.
4. Dael Wolfle, "How Much Research for a Dollar?" *Science 132* (1960): 517.
5. *Science and Common Sense* (New Haven, Conn.: Yale University Press, 1961).
6. See Figure 1, p. 196 in Rescher, *Scientific Progress.*
7. Figure 4, p. 201, ibid.
8. Figure 3, p. 199, ibid.
9. Thomas J. Allen, *Managing the Flow of Technology: Technology Transfer and the Dissemination of Technological Information* (Cambridge, Mass.: MIT Press, 1977).

PRESSURES ON THE STRUCTURE

6 IMPACT OF WORLD WAR II

Concern over national security has been the driving force for organizing massive approaches to difficult problems of technology during war and a solid source of support for R&D during peacetime. Military use of aircraft in World War I was a principal motivation for establishing the National Advisory Committee for Aeronautics (NACA) in 1915. This agency provided much of the technical foundation for the growing aircraft industry. The first large-scale electronic computer was developed for the Ordnance Department during World War II to carry out ballistic calculations. This was the Electronic Numerical Integrator and Calculator (ENIAC), designed and built at the University of Pennsylvania. And, of course, the Manhattan Project, which launched atomic bombs and nuclear power along with a new level of support for "big science" in the pursuit of basic nuclear physics research, came directly from World War II.

Every war in modern history appears to have pushed the technical base of mankind to a higher level of knowledge and sophistication. World War II changed the structure and role of the process for producing technical change. This was due partly to the specific experiences with integrating R&D into military strategy, and partly to the nature of the technical community that existed when the war ended.

The part played by science and technology in World War II influenced developments of the postwar industrial economies at three levels: (1) substantively in the advances in many technical disciplines,

117

(2) organizationally in the structures of R&D, and (3) psychologically in the attitudes with which society regarded the value and functions of technical activity.

Public attention focused on the substance of war-driven technology—electronics, materials, aircraft—and, above the rest, nuclear energy. Without these tangible outputs of such concentrated effort, no further impact would have resulted. Seeing the results of the effort brought a broadening awareness that a new potential had evolved for achieving economic and social objectives. The public was ready to rely more heavily on science and technology for nonmilitary goals, and there was a new organizational platform to accomplish these goals.

Two unique elements were fostered by World War II. First was the establishment of major technical programs requiring a broad spectrum of research and development activities in which the final product was part of planned military strategy. The same organization that initiated or at least monitored the technical progress converted the results to use. There was, in other words, an overall coherence among technical planning, R&D activity, and conversion to use. Despite some horror stories of how poorly this all worked on occasions, the successes in radar, in recoilless rifles, in the atomic bomb, and in many specific weapons developments are undeniable.

The second unique element was the nature of technical organizations outside the military after the war which could exploit the wartime experiences with organized R&D. The sector best prepared to capitalize on these experiences and with the greatest opportunities to do so was industrial research.

The organizational experiences can be considered in two categories. One is the specific methodology developed to plan and control the conduct of large complex systems. The other is the general approach to carry on a planned technical development in parallel with those other logistical and strategic functions—manufacturing capabilities, raw materials, personnel—that would all have to be coordinated on schedule to implement a specific objective.

The term *systems analysis* covers the particular methodologies used. It represents the formal approach to complex problems. A program that may sound purely technical (for example, developing titanium sheet for aircraft structures) or one that has a formidable technical-economic-resource objective (for example, developing an atomic bomb) is broken down into the vast array of projects, tasks, and subtasks that must be performed to complete the broad objective. Careful analysis must be performed in order to

1. Relate the timing and the results of each task to every other task
2. Determine the resources necessary for each task at a particular time
3. Assign priorities for completion of tasks
4. Prepare alternate options, or redundancies, for the most critical components
5. Identify the impact (in time, cost, resources) of difficulty or delay in one part of the system on the performance of the overall system

The process is complicated of course. A complete plan is not a list, but a series of maps plotted on a time chart. The related tasks are joined by lines intersecting at critical points on which many actions depend. This dictates priorities and allocation of resources.

Laying out an orderly list of tasks is hardly new. This has been the foundation for every military campaign and every major construction project throughout history. The new features about the systems analysis developed during World War II were

- Inclusion of R&D in the projects and tasks necessary to achieve specified objectives
- Willingness to evaluate the probability of success for each technical development
- Confidence in the technical foundation behind the broad program
- Management of technical uncertainty

Further, there existed at the end of the war a substantial and growing number of industrial research laboratories. All of them had participated in some aspect of war-related technical activities. They were on familiar terms with carrying out some portion of a complex system, and with the meshing of technical programs into a broader system of manufacture and use.

It is fairly obvious in hindsight that these characteristics constitute the essence of modern organized industrial research. The features that constituted the newer aspects of systems analysis sound like the philosophy we have come to expect from any progressive technically based corporation today. They would be the principal components of the discipline that we call "research management," and indeed make up the contents of textbooks and seminars on that subject.

The significance of the military experience for the growth of industrial research was not at all obvious. Many individuals were well aware of the methodologies, the implications, and the applicability to complex technical programs in any area, public or private. But these individuals were research executives, program managers, whose authority was limited to the conduct of R&D. The integration of technical programs with corporate objectives required the authority of higher corporate management.

Two critical elements differentiate military use of science and technology from civilian use: namely, marketing the outputs and financial return on the inputs of R&D. These elements are missing from the normal military criteria, although they are present indirectly when choosing among options. There would be little reason to expect the military experience to be transferable to civilian programs in view of these differences.

The stimulus for absorbing those experiences into industry was psychological. Management had just witnessed the conversion of technical advances to dramatic practical applications in communications, transportation, and energy. The technical sophistication available for use in the emerging peacetime economy was vastly superior to the level in the civilian economy of the late 1930s. Not only was there a new array of technical resources to draw upon, there was evidence that we could organize resources to focus activity on desired objectives with reasonable confidence of achieving practical results.

A general expansion of industrial research took place in the 1950s and 1960s, including the establishment of new laboratories. The willingness—in fact the enthusiasm—of top management to embark on plans for growth in new products and processes that required the successful pursuit of major R&D efforts was a direct by-product of the confidence developed from the successful uses of technology in World War II. There was a sense that science and technology provided the key to improved quality of life, that R&D was the basis for corporate growth.

The vigor and initiatives reflected in the expansion of research laboratories in industry led to its share of errors. There were misunderstandings about the nature of research. Some laboratories were later closed or reduced sharply in size. Nevertheless, modern industrial research matured in the enthusiasm of the 1950s and 1960s. It was encouraged by the successes of technology in World War II and it had the opportunity and capability to adapt the relevant methodologies.

EXPERIENCES WITH TECHNICAL COMPLEXITY

The development of laboratories within the structure of the corporation provided the basis for effectively continuing the accelerated growth in technical resources. The ability to tap directly related financial resources, the more rapid and focused feedback between technical activity and use, and the combination of disciplines for mission-oriented programs were all critical elements in advancing the technical process.

In fact, while the early part of the twentieth century witnessed the spread of research laboratories to an increasing number of companies as well as a steady growth in personnel within the early laboratories, the nature of industrial research organizations was not very complex by today's standards. A single central laboratory interacted with one or several manufacturing divisions. The corporations were also less complex, so that the interests of each laboratory, while broad and covering many disciplines, could at least be stated fairly simply.

Industrial research matured quickly in the two and a half decades after World War II. The initial emphasis was on the internal operations of a laboratory—treatment of technical personnel, design of laboratories, accounting systems for R&D, organization of research. This quickly expanded into attention to the relationships between R&D and other corporate functions—marketing, manufacturing, finance. By the mid-1960s, senior R&D managers were looking increasingly at the impact of activities outside the corporations, at government policies, growth of R&D abroad, and shifting patterns of international trade.

All of this activity simply defines the increased magnitude and complexity that accompanied industrial research in the years following World War II. Obviously, these characteristics mirrored the increased magnitude and complexity of the industrial expansion itself.

The impact of the war on the structure and role of R&D in industry was therefore important in two developments:

1. Increasing the confidence of industrial top management in the ability of R&D to provide the necessary products and processes that fed the expansion

2. Increasing flexibility and capability of industrial research management to transform their organizations so as to conduct more complex, more coordinated, more scheduled programs

TECHNOLOGY AND STRATEGIC PLANNING

The enlarged scope of industrial research after World War II paved the way for closer and broader relations between technical change and strategic corporate planning which characterize the modern industrial corporation. The evolution of industrial research would surely have led to this status over a longer period of time. Nevertheless, the war serves as a clear boundary between the functioning and significance of industrial research as it was conducted in the first half of the twentieth century and as it exists today to facilitate our entry into the twenty-first century.

Placing a research laboratory within a corporation was a vital step in the process of technical change. Advancing the role of research so that it is a key factor in the planning of the corporation, with the chief technical executive part of the senior management, increases the effectiveness of the process and thus its contributions to society.

The most sophisticated companies, with the longest traditions of organized research, began to develop a pattern of integrating these R&D advances with marketing and financial planning well before World War II. The development of nylon at duPont, the research in high-pressure physics that led to synthetic diamonds at General Electric, the many advances in communications and telephony within the Bell system were all evidence of how management was learning to accept the progress from R&D in terms of the steps necessary to convert that progress into commercial use. This became a more familiar process for those companies with longer experience in growth from technology.

For most American and European companies, the R&D organization was a valuable adjunct to corporate operations. It was a source of continuous improvements in productivity, and it could be expected increasingly to come forth with the technical base for new products and perhaps new business. It provided a source for corporate growth. However, it was not typically an active component in corporate strategic planning. This changed steadily after World War II.

Industrial research did not originate in the United States, the German chemical industry generally being credited with the first industrial laboratories in the late nineteenth century. But there is one uniquely American contribution to industrial research that evolved throughout U.S. industry in the 1950s. This was the deliberate, planned use of R&D as a mechanism for corporate growth.

This was impressed upon me in 1959 when my position as director of research for American Machine & Foundry Company (AMF) required the establishment of a research laboratory in England. Candidates for the post of director of the planned laboratory were attracted by the opportunity the American company offered them to interact with members of management or, in current language, to be a "member of the team." Procedures that I took for granted, such as access to top officers of the corporation or participation in meetings with the chairman, were not common in European industry.

The readiness of top management to base corporate growth strategy on specific technical advances by the corporate R&D organization arose from the wartime successes in planning and executing technical developments. The value of industrial research was established and the probability of deriving economic return from R&D investments was improved. This was an important factor in the growth of "big" industrial research.

The evolution was, in hindsight, inevitable, but World War II provided corporate management with the confidence to expand and elevate industrial research. This was an important head start for U.S. industry in the postwar world. Today, of course, it is accepted procedure within industries throughout the Western world and Japan, but there was a gap in management approaches that may have been 10 or 15 years in duration.

The perception of technical self-sufficiency throughout this expansionary period, roughly 1950 to 1980, is critical to our understanding of the technical enterprise today. To the corporation, technical self-sufficiency means that the corporation has the resources to generate the technical base for the products, processes, and services that will constitute the business operations at sometime in the future. The technical advances sought must be within the financial resources of the corporation and timing of the many discrete technical studies must be in synchrony with the corporate strategic plan. There must be a reservoir of competent technical personnel at all levels. The level of knowledge in related fields and the necessary advances in instrumentation must be available or within the resources of the corporation to provide. In other words, technical self-sufficiency depends on a pool of adequate funds, knowledge, instrumentation, and personnel. By definition, a well-managed corporation is always technically self-sufficient, since it will not adopt a business plan for which it cannot provide the necessary technical base.

In the postwar period a very high rate of industrial growth was possible for which the individual companies were indeed technically self-sufficient. The conditions that existed in the U.S. in 1950 and beyond provided all the necessary components. There was a very considerable reservoir of science and technology in terms of the advances in electronics, materials, instrumentation, mechanics, and indeed all the technical disciplines. There was a strong and growing university base that provided increasingly better trained technical graduates. There was a massive influx of federal government funds to support defense, nuclear energy, space, and health that advanced science and technology in many fields simultaneously, so that a program in one area could draw upon advances in others. And, from the corporate view, there were expanding unfilled markets in the United States and around the world that could provide the necessary financial resources.

The increased corporate resources and the deliberate use of industrial research for corporate growth were mutually reinforcing. Increased industrial research, backed by growing corporate resources, permitted technical advances in every area. In fact, as corporate resources grew, broader and more complex technical problems could be approached and completed.

Inevitably, the restrictions imposed by the requirements for technical progress must apply. For at least 20 years, from 1950 to 1970, the increase of corporate resources and of technical capabilities everywhere appeared to occur at a rate faster than the increase in resources necessary to advance different fields of science. These rates began to reverse in some fields by the late 1970s, as technical advances required increases in cost and personnel at a greater rate than the increases in corporate resources. In that sense, the important phase of technical self-sufficiency following World War II probably began to decline by 1980.

A LOOK BACK

The unspoken sense of confidence with which industrial management and research executives approached the postwar period was a factor that can hardly be overestimated. In the 1970s, it was common to approach all sorts of societal problems with the phrase, "If we can put a man on the moon, why can't we . . . ," without sufficient attention to the political and sociological factors that are present in improving cities, transportation, and developing countries. But the concerns of industry in the 1950s and 1960s were with technology and economics, which, while difficult, had far fewer restrictions than the political questions present in the public sector.

Thus, the confidence, the methodologies, and the familiarity with organized research were built into the ambitions and the thoughts of U.S. chief executives after the war. The growth of AMF, for example, from a modest-size manufacturer of industrial machinery to a major conglomerate in electromechanical equipment, nuclear research reactors, relays and controls, recreational equipment, and other areas had one common key element. The chairman and chief executive officer, Morehead Patterson, had a deep conviction that the new world of science and technology created a new world of opportunity for industry. Anything that was desirable from marketing and economic criteria was possible technically. That mood was characteristic.

Larger corporations with older R&D organizations had developed this attitude before the war. Nevertheless, research managers at GE, duPont, Bell Laboratories, and others all experienced the same sense of top management willingness to plan *with* their technical executives because of the broad acceptance of technical change as an instrument of opportunity and growth.

Today, we have a body of experience that may be called "research management" or perhaps the "management of technology." Certain aspects of this are fairly recent experiences, for example, start-up ventures within a corporation. Nevertheless, the greater part of our current grasp of managing organized research for the purposes of economic growth evolved in the postwar period and grew out of wartime experiences. It was based on technical advances, on methodology, and on psychology.

7 THE TECHNICAL ENTERPRISE OF THE 1980s

The exuberant growth of R&D in all sectors during the quarter century after World War II was characterized partly by the increase in money and people, partly by new facilities and new research organizations, and partly by new relationships. This all occurred in the midst of comparable growth and shifts throughout all institutions of society, with increasing interactions among technical activities, international trade, market growth, and government policies. The question of which is cause and which is effect during a period of rapid change is perhaps unanswerable and very likely academic.

What is clear is that patterns were set during that period for the worldwide structure of R&D that has been referred to throughout this book as the "technical enterprise," a structure that differs in its functions and interrelationships from the structure of technical activity that existed in any previous period of civilization. The form of that structure was reasonably clear by the late 1970s. It represented both a consolidation of the growth that occurred during the postwar years and, at least in the United States, a maturing of the relations between government and industry, and between university and industry. Perhaps most important, the framework of multinational corporations was put in place, and a broad internationalization of business was well underway.

NEW PLAYERS ON THE WORLD
TECHNICAL STAGE

By the late 1970s, U.S. technical activity in all sectors had become the major force for scientific and technical progress in the world. University research was the foundation for basic science and engineering. The federal government provided a funding source for basic research, including the massive facilities needed to advance nuclear physics, and conducted major R&D activities for the public sector, such as space exploration. Industrial research functioned as the principal engine for the production and conversion of technical advances to economic use, and increasingly as the driving force for stimulating R&D throughout the entire system.

The magnitude of the U.S. technical effort was only one aspect of the domestic technical enterprise. Understanding of the role played by each sector improved and the mechanisms to enhance the productivity of the total R&D system grew. University-industry relations were strengthened during the 1970s. Government policies regarding R&D expenditures had proved highly effective up to 1970 by focusing on areas where the government was the final customer. There were some missteps during the 1970s when the government turned toward programs intended to be implemented within the civilian sector, for example, in energy and in the Cooperative Automotive Research Program (CARP), but these difficulties were recognized by 1980, and smoother relations followed.

The technical enterprise within the United States was performing well by 1980. There was no essential problem with the continuing generation of technical change. There is none today.

There were, of course, very considerable changes in the rest of the industrialized world in the postwar period with regard to both industrial growth and increased capacity for the conduct of R&D. Although Japan and the European countries started from a smaller base largely damaged by the war, their growth was in many cases much more rapid than in the United States.

The resurgence of the technical bases in Europe and Japan which occurred in the 20 to 30 years after World War II created a new foundation for the overall technical enterprise of today. Certain aspects of this growth are relevant to our general theme of the conditions surrounding technical progress.

The dynamic factor in the growth of technical activity outside the United States was, again, industrial research. Certain industries spend more on R&D than others, and a minimum size is required to mount a broad effort, particularly one that includes long-term research. Industry is likely to invest in R&D only in so far as the R&D is converted successfully into products and processes *and* there is a market which can provide the necessary revenue. The size of the R&D effort depends on the size of the market available to the corporation.

In the postwar period, there were three sources of markets for the European and Japanese companies: government procurement, a rising domestic market, and the United States.

Government procurement is always an important factor, but there are two limitations. First, it tends to focus on specific products related to defense, transportation, and communications (the national Postal, Telegraph, and Telephone services, or PTTs are important purchasers of electronic equipment). Second, since these areas are government monopolies, the technical approaches and the resulting manufacturing costs are not necessarily subject to the competitive disciplines of selling to a wider market. Thus, procurement will provide revenue and permit support for R&D, but it does not automatically provide thr capacity to compete successfully in the international marketplace.

Domestic markets in Europe required some time to build up after the war. The needs were there, but the economies had to be stabilized. More significant was the small size of the domestic market in the individual countries. Europe was a large market, but Belgium, England, and Holland were not. The eventual formation of the Common Market helped in theory, but the desire of each country to preserve its own principal corporations in each major industry set limits to achieving the advantages of larger operations through wider market access.

I remember attending a conference on the "technology gap" between Europe and the United States at the inauguration of the present facilities of the National Bureau of Standards in November 1966. Seated next to me was H.B.G. Casimir of the Philips Company, Netherlands (a friend and colleague). When solutions were being considered, he rose and said: "Set up a tariff and customs office at the borders of each of your States. Preferably, have the citizens of each State speak a different language. The technology gap will disappear overnight!"

Thus, the major available competitive market for European and Japanese industries after the war was the United States. The quality of the scientific and technical base overseas was excellent, though depleted and somewhat disorganized by the war. It was relatively straightforward to proceed with the tasks of strengthening research in the public sector—universities and government laboratories. Public policy and intent were clear. The build-up of industrial research, however, required assurance of adequate size of operations to justify the necessary investments in R&D. This meant access to U.S. markets, either directly or indirectly via some form of arrangement with U.S. corporations, in a joint venture or working through a subsidiary.

When we look at the strong technical organizations today throughout Europe and Japan, integrated within large and powerful multinational corporations, it is difficult to appreciate the step-by-step efforts that were necessary from 1945 to, say, 1960 to build a stable foundation for further expansion. A technical organization was necessary to adapt the technical advances that had taken place in the war years in order to design competitive products and processes. Factories and equipment had to be prepared to manufacture the products and use the processes. The outputs had to be sold. Funding had to flow to expanded R&D that could broaden the technical base for the next generation of products. This was very much a bootstrap process in the years just after the war.

The rate of growth for industrial research outside the United States was accelerated by the fact that U.S. markets, both consumer and industrial, were open to corporations from the other industrialized countries. So, also, was U.S. science and technology. Scientific advances were disseminated through traditional professional channels—journals, meetings, reports. Technology was available through licensing. The technical interactions between U.S. companies and their counterparts in other countries set the framework for more extensive linkages that we see today, and will discuss more fully in the following chapters.

The United States provided a technological head start to bridge the disruptions of World War II. Nevertheless, the industrial expansion of Europe and Japan was accompanied by a comparable expansion of the technical base in industry. The investment decisions and the market strategies, both requiring greater technical efforts, were not U.S. decisions.

The results are well known. From 1960 on, Europe and Japan became major forces in industry generally, and in industrial research particularly. In electronics, the strength of Japanese companies became an awesome force, while Philips built one of the finest laboratories existing in industry anywhere. In chemistry, Hoechst and Bayer, ICI, Akzo, Montedison all were in the first rank of R&D capabilities. In other industry sectors, there flourished the great laboratories of Shell, the metallurgical facilities of Pechiney, the electrical advances of ASEA and L.M. Ericsson of Sweden and, of course, the famous Swiss pharmaceutical companies, Hoffmann La Roche, Ciba-Geigy, and Sandoz.

The United States observed this technical and industrial growth outside its borders first with friendly interest, then as opportunity, and finally, by the late 1970s, with growing apprehension in many quarters. The world was no longer a U.S. pond technologically, if indeed it ever had been. The rate of U.S. technical activity compared to that of the Federal Republic of Germany or Japan was dropping sharply, not surprising given the modest effort in those countries as of 1950. Other countries showed more rapid increases in indexes of technical intensity such as R&D expenditures as a percentage of gross national product, or number of technical people per 10,000 persons employed. The number of patents filed in the U.S. by foreign nationals rose, and the U.S. share of high-technology products in international trade declined.

None of this should have been unexpected, and the U.S. leadership in total resources devoted to R&D in actual dollars appeared to have increased by 1980 compared with any country except for Japan. The perception, however, was that the U.S. capability to remain strong technically, at least in terms of converting technical advances to products competitive in international trade, was slipping.

A critical stimulus for these perceptions was, in fact, derived from a series of difficulties in the late 1970s. The United States was experiencing a slowdown in economic activity and a rise in inflation. Japanese consumer electronic products had effectively taken those markets from U.S. firms. The German "economic miracle" was demonstrating its capacity to convert technical advances to machine tools that could compete throughout the world. Foreign automobiles began to claim a substantial share of U.S. markets.

All of this was summed up as a concern with innovation and, more specifically, with international competitiveness. These concerns involve the capacity for generating science and technology and the ability to put technical advances to economic use. There is no basis for questioning the U.S. capacity for creating technical change. If the

U.S. output of new products or U.S. performance in international trade is considered unsatisfactory, then the conditions for converting technical change into competitive industrial output must be examined. This is what emerged, in part, from the study on innovation initiated by the White House and the Department of Commerce in the late 1970s.[1]

Thus, as of the 1980s, there was a strong technical base outside the United States, and an uneasiness in the United States that perhaps our own technical base was somehow inadequate for the competitive years ahead. Such uneasiness was, and is, unjustified. The comparison with other countries can be considered on three levels. The ratio of our performance to others may well decrease as activity overseas increases. The technical gap between the United States and others is very likely increasing. And the absolute level of U.S. technical effort is steadily increasing.

It is a fact that others outside the United States are technically strong today. The world technical enterprise now has many players. This creates competition, but it provides greater thrust for advances in science and technology.

U.S. RESEARCH EXPANDS OVERSEAS

A phenomenon of the postwar years was the establishment of research laboratories in Europe and, to a lesser degree, in Japan by U.S. corporations. The reasons were varied, the successes mixed, and the conditions unique to each. They represent an important set of communication mechanisms and personal interactions that provided the starting point for eventual technical and commercial relationships. These go well beyond the original objectives for the establishment of laboratories abroad. Some of the early establishments included IBM and RCA in Zurich, Union Carbide in Brussels, American Cyanamid just outside Geneva.

Several reasons have been given for the establishment of these laboratories. Mentioned most frequently was the desire to have a "window on foreign science." It should be remembered that prior to 1940 there was a tradition of reliance on Europe by the United States for advances in basic science. A second point often heard was the lower cost of conducting R&D abroad, a perception that was quite erroneous. Each item of direct cost was less, but total R&D

productivity and effectiveness in converting results for use by U.S. corporations were rarely taken into account in that comparison. Other objectives included access to special skills not easily available in the United States, developing new sources of technical concepts, or simply establishing on an international stage the corporation's concern with science and technology.

If these reasons were valid, it is interesting that many corporations decided not to set up foreign laboratories and others closed their European laboratories within a few years.

The successful laboratories were set up by and large by companies with a manufacturing operation in the same country. From the view of research management, this correlation was to be expected. An industrial research laboratory does not exist in a vacuum. The interaction with a manufacturing arm provides some criterion for selectivity among the range of possible technical projects. It allows for feedback between technical approaches and manufacturing conditions. As a practical matter, the operating unit was often able to provide administrative services for the laboratory, particularly important in the early stages. Further, the operating unit may derive some immediate value from the laboratory's output.

These technical outposts of U.S. companies did indeed form important linkages with technical activity abroad. These linkages were valuable in facilitating the technology flow and commercial relationships which we see today within the international technical enterprise. The scientists and engineers within laboratories were part of the rapidly growing technical community abroad. More than that, because there often was a relationship with a manufacturing operation, they were part of the industrial community. They were associated not only with R&D, but with products and processes. Moreover, a laboratory in one country which is part of an industrial organization headquartered in another is an important component in the operation of the worldwide technical enterprise. This is just as true for the scattered laboratories of Philips, Hoechst, and Hoffman La Roche as it is for IBM, RCA, and ITT.

In the 1950s and 1960s, the establishment of U.S. laboratories abroad served the purpose of transferring practical technical know-how to Europe and Japan as they rebuilt after the war. Besides transferring technology through personnel, multinational corporations transfer technology through cross-licensing and common products and processes. If a corporate R&D map of the world were drawn, with a dot for every laboratory and the dots connected for Philips, ITT, Shell, the Swiss pharmaceutical firms, the German chemical companies, and all the rest, we could see clearly the pattern of technology flow through the individual loops and across national boundaries.

In addition to developing their internal resources for international R&D, companies are establishing ties with relevant R&D resources throughout the world. The most accessible international resources are universities. However, there are also private research institutions and government-sponsored organizations which can develop ties with companies headquartered in other countries. For example, the TNO in Holland accepts research contracts with foreign companies. This is a very large quasi-governmental research organization, and thus quite different in its public connections and obligations from such private contract research organizations as Battelle, SRI International, or Arthur D. Little.

An American corporation develops relations with a French or German university usually through a subsidiary or division in France or Germany. IBM, for example, has a European research network connecting its European laboratories with European universities. Since its operations in each country are corporate citizens of that country, the university relationships there are analogous to its relationship with MIT. Foreign-owned multinationals have reached out to American universities as well. The arrangement involving the most money thus far is probably the agreement between Hoechst of Germany and Massachusetts General Hospital, the research arm of the Harvard Medical School. This was a 10-year agreement for $50 million beginning in 1981. It included research sponsorship in the field of molecular biology, with certain rights for Hoechst, including assignment of Hoechst personnel at Massachusetts General. This agreement stimulated some objections in the United States from people concerned with foreign companies "walking off" with the fruits of U.S. research.[2]

There has been similar debate recently on the growth in funding of U.S. university research by Japanese companies.[3] Among the large grants is $5 million by Toshiba to support digital radiography at the University of Arizona and $600,000 from Hitachi to the University of California (Irvine) for research in biochemistry. MIT, Tulane, and Georgia Institute of Technology have pushed vigorously for Japanese support. Of those few universities which have stayed away from funding from Japanese companies, Carnegie-Mellon University has taken a clear position with regard to its research in robotics and artificial intelligence. Provost A.G. Jordan has stated, "There is some concern on our part that this would be a transfer of technology to Japan which we should avoid." In an editorial, *Business Week* observed that "a lot of businessmen are surprised and dismayed" at Japanese funding of basic research in U.S. universities.[4]

Given the extent of ties between U.S. corporations and universities in other countries, these objections to support by foreign companies are at best inconsistent. More seriously, they demonstrate a fundamental misunderstanding of the nature of the technical enterprise from which all countries have benefited, the United States being probably the greatest beneficiary.

Each time a corporation develops a satisfactory working relationship with a university, particularly one in another country, the overall process of technical change is broadened and expedited. Obviously, the amount of research is increased. Further, technology flow is facilitated. Even with some restrictions on access to the results, the arrangement is a two-way street. All parties are better informed, and communication is increased.

Such corporate relationships with external resources in other countries form important branches of the technical enterprise. The stimulus to the system should benefit all participants.

OTHER INTERNATIONAL LINKAGES

Numerous other mechanisms serve to create technical relationships on an international scale. Most of these are between governments, some are between companies, a few involve both. Such ties address a public or institutional objective, not the specific commercial objectives of corporations or the traditional mechanisms for exchanges in basic science.

There has been a considerable increase in scientific and technological agreements among governments. The bulk of these are bilateral, between the United States and France, Israel, and Japan respectively, for example. These tend to encourage basic scientific research through cooperative projects, exchanges of personnel, workshops, and visits. A number of agreements are multilateral, involving many countries. These are usually mission-oriented, dealing with such problems as energy conversion and generation or environmental pollution. Occasionally, such multilateral agreements set up a planned technical program, with each member conducting some component of the necessary R&D. An excellent example of this is the International Energy Agency (IEA), with many active research programs. For the most part, however, these agreements serve more to stimulate R&D and increase the flow of science and technology than to carry out R&D.

Government agreements draw upon researchers and ongoing projects from all sectors. Agreements emphasizing basic research involve university researchers. The mission-oriented agreements often involve researchers from government or industry. This mechanism serves, therefore, to develop cooperation among all sectors on an international basis. It expedites diffusion of technical progress beyond traditional professional channels.

Other forms of institutional and industrial linkages internationally strengthen the infrastructure of science and technology. Many of these have to do with the establishment of technical standards, a function critical for international trade and communications. Standards are set officially by governments, who draw upon industry representatives for advice. This occurs in telecommunications through the International Telecommunications Union (ITU) and in transportation through the International Civil Aviation Organization (ICAO).

A handful of international agreements involve the private sector as the key participants, although they were officially initiated by governments. One good example is the Experimental Safety Vehicle (ESV) program. It began in the early 1970s, and in 1974 was renamed the Research Safety Vehicle (RSV), Automobile manufacturers in seven countries, including the United States, participated. A principal contribution was the exchange of data on technical advances related to safety technology, fuel economy, and other subjects where one could separate highly proprietary approaches from data that would be useful to all.

Finally, there are a great many programs under the sponsorship of international agencies, most of which are concerned with the developing countries. These countries are not yet significant contributors to the worldwide technical enterprise. Yet considerable effort is being exerted to develop a technical infrastructure in these countries and to establish mechanisms for the transfer of science and technology to developing countries from the developed industrialized nations. The bulk of these programs are supported by a range of international agencies. The total budgets for science and technology of those agencies for the year 1978-79 came to $813 million, of which $335 million was in the regular budgets and $477 million in extrabudgetary items.[5] Almost half of the total, $403 million, came from the World Health Organization (WHO). It also includes the International Atomic Energy Agency (IAEA) with $44 million, the World Meteorological Organization (WMO) with $20 million, the International Labor Organization (ILO) with $68 million, and the Food and Agriculture Organization (FAO) with $44 million.

The bulk of these activities are within the United Nations system. They emphasize public sector science and technology and strengthening the technical infrastructure. Some of the technical activities of these international agencies are intended to strengthen the industrial base (that is the mission of the United Nations Industrial Development Organization and the United Nations Development Program), but these constitute about 5 percent of the total budget for the group of agencies.

The technical efforts of those international agencies concerned primarily with developing countries serve largely as one-way branches from the main technical enterprise. More stimulating to economic growth in the less developed nations would be through technology transfer by multinational companies establishing subsidiaries or joint ventures there. That is undoubtedly the most effective system for developing skills needed to integrate technical advances into industrial operations and to provide the familiarity with industrial research management which can stimulate the overall technical economic process.

FRAMEWORK OF THE TECHNICAL ENTERPRISE

The technical enterprise of the 1980s is a massive effort that must be separated into the principal sectors—university, government, industry—and must be viewed on a global basis. Every location which conducts R&D is, in principle, communicating with all other locations through traditional professional channels—meetings, journals, reports, visits. It is as if each location is floating in an ocean of science and technology which it absorbs and into which its own output flows. The water level rises constantly.

As we view the system more closely, a pattern of linkages appear that is more directional. Technical ties connect each major company and government laboratory to selected universities. In each country, government laboratories in mission-oriented fields develop linkages with industry laboratories relevant to those fields, such as energy, transportation, mining, and, in the United States, a broad set of relationships between corporations and the National Bureau of Standards (NBS).

Then there are networks that bring together clusters of activity to form closed loops, some large and some limited. Each major technically based corporation has its internal network connecting its divisional laboratories and its central research establishment. Where the corporation is multinational, the network extends across national boundaries. The growing number of technical areas that provide an opportunity for constructive action by governments acting cooperatively form their own networks of international agreements. The largest network of all is the defense network, which links together many R&D institutions, public and private and is superimposed on all the others.

The technical enterprise contains enormous numbers of intersections. The loops intersect. Each location can be a part of several major external networks, part of an internal network, and can maintain many one-to-one ties with other institutions. Thomas Allen at MIT has popularized the concept of "gatekeepers" in a laboratory or in a country who are the principal absorbers of external science and technology. His work deals with individuals. I suggest that the concept is far more general. Particular companies (such as GE), particular universities (such as MIT), and particular government laboratories (such as NBS) are the gatekeepers for the technical enterprise as a whole. The phrase refers not to what R&D is generated, though that is closely related, but to what is communicated. Those locations which belong to many networks and have many relationships provide a much broader function for the entire technical enterprise than simply advancing the objectives of the individual locations. They become transfer points from one network to another, from one linkage to another.

The tens of thousands of locations generating R&D constitute the body of the technical enterprise. The innumerable linkages and networks within the system form the veins and arteries. There is constant flow, constant absorption, constant regeneration. The health of the whole system lies not only in the ability of one location to generate R&D, but on the capacity of the system for the flow and absorption of science and technology. An obstacle or restriction in flow will cause new networks to form around it, or will weaken the overall performance.

Basically, the technical enterprise is healthy. Its size and diversity provide flexibility and momentum. It is, however, a growing organism. What we must consider as we look ahead is how continuing growth can affect individual components of the system and what conditions are most helpful to permit orderly growth to occur and still maintain the overall health of the entire technical enterprise.

NOTES TO CHAPTER 7

1. *Domestic Policy Review of Industrial Innovation*, U.S. Department of Commerce, Office of Assistant Secretary for Productivity, Technology and Innovation, May 1978.

2. Soon after the Hoechst agreement was announced, I was visiting in Bonn with a colleague who was a Ministerialdirektor of the Bundesminister für Forschung und Technologie, the German research ministry. He commented, "If you think Americans are disturbed, you should hear the opinions of the German universities!"

3. Japan Is Buying Its Way into U.S. University Labs," *Business Week*, September 24, 1984, p. 72.

4. "Paying the Piper for U.S. Research," *Business Week*, October 1, 1984, p. 134.

5. K.H. Standke, "Scope and Function of International Technical Arguments," in "Industrial Productivity and International Technical Cooperation," edited by H.I. Fusfeld and C.S. Haklisch (New York: Pergamon Press, 1982).

8 THE TECHNICAL IMPERATIVE

Pressures have developed in the technical enterprise from the interaction of two processes. One is the continuing trend of scientific and technical progress whereby each new technical advance requires more resources than the preceding one and the amount of resources devoted to R&D must increase at an exponential rate in order to produce significant technical advances at a constant rate. The second process is the increasing dependence of industry on technical change and on its own capacity to produce change, that is, industrial research. To a great extent, this dependence follows from the success of modern industrial research. It is reflected in the shorter life cycles of new products in the electronics industry; in the steady growth of biotechnology applications in established industries such as foods, chemicals, petroleum, and pharmaceuticals; in the drive for increased productivity through the adaptation of microelectronics, controls, and robotics in all manufacturing; and in the constant pressure of technical advances throughout the world in process industries such as steel, in transportation, and in new materials.

Before the introduction of industrial research, there was a fairly clear distinction between the generators of technical advances and the users. Academic researchers, individual inventors, small laboratories were generators, not users. Government, primarily the military, supported some programs in which the government was both the generator and user. Much of this R&D enriched science and technology in many fields, such as chemistry, mechanics, thermodynamics, and metallurgy. But the government as a user focused on those applications specific to government purposes. The conversion of technical advances to economic use in society was performed largely by organizations that did not generate those advances.

This separation of user and generator was inefficient. It required a longer time to diffuse knowledge and transfer know-how, and it deprived manufacturing, the ultimate user, of feedback.

Everything is indeed relative. These transfer processes in the eighteenth and nineteenth centuries were "inefficient" compared to today. Still, they represented enormous improvements over the more random process of technical change existing before that. This is summed up by Jean-Jacques Salomon (at the Conservatoire National des Arts at Métiers, Paris), a distinguished scholar and historian in the field of science policy:

> During the 18th century, the rate of technical progress was unquestionably speeding up, but one cannot justifiably speak of a technical revolution. One can only fully perceive (and measure) the acceleration of technical progress from the end of the 19th century. Up to the Middle Ages technical progress was measured in millenia, and only then did its pace begin to accelerate, first from century to century, then half-century to half-century, and in our time from quarter-century to decade.[1]

The introduction of industrial research served to increase resources and thus accelerate scientific and technological progress. The impact on the corporation was significant. Industrial research became a conscious mechanism for corporate growth, integrated with marketing, finance, and manufacturing in strategic planning. The R&D function was more crucial in some industries than others. Nevertheless, the principal characteristic of the modern technically based corporation, influenced strongly by its experience with science and technology during World War II, was its willingness to base strategic plans for future growth increasingly on the performance of R&D not yet completed.

This is a significantly different pattern of behavior for modern industry than existed before industrial research evolved.[2] In earlier times industry accepted and absorbed technical advances generated by others but did not plan for such change or control it.

Industry today does plan for technical change and with confidence when it has control over adequate R&D resources. For the first time in history, the largest user of science and technology has become the largest generator. By 1980 industry expenditures for R&D exceeded those of the federal government in the United States and in every major European country and Japan except France and England, where industry funded from 40 to 43 percent of the total national R&D.[3]

The most obvious consequence of this user-generator status has been improved effectiveness of the total technical enterprise, resulting from the greater feedback between the sources of technical advances and the user. Another important result has been the world-wide increase in resources devoted to R&D. University and government R&D expanded first to satisfy public objectives in defense, health, and space and more recently to support economic growth. As European and Japanese companies emerged after World War II to enter world markets influenced by U.S. industrial research, a substantial build-up of their own internal R&D organizations was inevitable and rapid.

Clearly, the greater resources and effectiveness of the technical enterprise have promoted technical progress in all fields. But since each next advance has always proven to require more resources than the last, a time is bound to come when demand will exceed the resources available for R&D. The impact on the individual corporation can be considerable.

INDUSTRY'S CAPABILITY SQUEEZE

Corporation managers must balance their corporations' needs for technical change against their financial capacity to provide for those needs. Maintaining product quality and developing new products are normally within the capacity of the internal R&D organization with some outside consulting or contracting. (The corporation that lacks such capacity has strategic difficulties much more immediate and fundamental than the kind of concern here. This may be the situation today in the mining industry due to the restrictions of world markets and poor cash flow.)

In the more general situation, the corporation develops a plan for growth, or even for holding its own, that requires the conduct of a major technical program. The program may require (1) expansion of the corporation's own R&D organization, (2) work in new technical fields not within the present competence of the internal organization, (3) a great concentration of technical resources, or (4) any combination of these.

Until recently, the majority of business plans that depended on technical change were within the competence of growing industrial research laboratories. And industries experiencing economic prosperity and growth in the boom years after World War II could afford technological change.

On the heels of the accelerated technology push of World War II, further advances were possible with practical resources. The broadening reservoir of science and technology opened many fields and subfields, so that a corporation could identify growth opportunities where the technical advances called for were within its capabilities.

The first turning point in industrial research was the willingness of corporations, following World War II, to plan for growth based on technical change. Now we are approaching a second turning point: the declining technical self-sufficiency of individual corporations. Frequently, the technical resources needed to accomplish strategic growth are outside the R&D organization, and increasingly they may be beyond the capacity of the individual corporation to coordinate or control.

Consider the situation of Eastman Kodak, one of America's great technically based corporations. There is a high rate of technical change in the photographic field, including the increased use of advanced electronics, not the traditional base of Kodak's technology. Kay R. Whitmore, president of Kodak, summarized the situation as follows: "We've come out of an environment where we were the single world leader, we had a technology that nobody else could really match, and we were able to dominate that field. The world doesn't allow companies to do that anymore."[4]

Whether corporate technical self-sufficiency is possible depends on the following factors.

Share of Technical Effort in a Given Field. The R&D effort of a single corporation in a technical area is, with rare exceptions, a small fraction of the total R&D effort devoted to that area within the technical enterprise. As all sectors increase technical activity, this fraction decreases. The impact is particularly noticeable to U.S. corporations today in comparison with the situation 10 or 20 years ago because of the resurgence of European and Japanese industry. As an example,

Japanese companies developed only 1 percent of new drugs in the 1970s, but are responsible for 20 percent today.[5]

Number of Technical Fields Relevant to Corporate Growth. Continuing technical progress opens new branches of knowledge and creates new interfaces between existing disciplines. Chemical companies are turning to biotechnology, automotive and process industries use robotics and microprocessors, and medical instrumentation combines optics, mechanics, electronics, and biology. As products and processes become more sophisticated and represent more substantial technical change, they tend to require contributions from a larger number of scientific and engineering specialties. It is a different era from just 40 or so years ago, when the American Institute of Physics embarked on fund-raising for a headquarters building. In a letter on file at the AIP, the senior technical officer of a major metals company declined to contribute since "we are engaged in metallurgical research, and therefore have little interaction with physics." That was before jet engines, ultrasonic testing, composite materials, and nuclear reactors!

Requirements for Significant Technical Advances. In industries which are characterized by a high degree of technical change, as well as in industries which have experienced minimal technical change, business growth frequently requires a massive technical step forward. Development of the so-called fifth generation computer or the 256K chip is in this category. Laser devices, photographic imaging, medical instrumentation, biotechnology, catalysis are all areas where the emerging products are based on pushing the technical frontiers of those fields. Wholly new approaches for mining, for treating chemical waste, for continuous casting of all common metals, for reducing friction in mechanical systems are less glamorous but equally complex or massive in their technical requirements. More and more companies identify an important growth opportunity, perhaps a necessity, that is dependent upon a very significant technical advance which they are unlikely to achieve alone, as individual firms.

A corporation can remain technically self-sufficient by increasing its technical resources or by decreasing its needs. Thus, technical self-sufficiency depends upon the growth strategy of the individual firm. A company which identifies and pursues a niche in a market does not require the same technical resources as one which opens new markets. A corporation in a capital-intensive process industry with a poor cash flow or declining market may focus on improving present processes to lower costs, a much smaller technical demand than the development of a radically new process. Most corporations grow and prosper by market development and productivity improvements based upon very modest technical change. There may be substantial investments in engineering and in equipment, but very little technical uncertainty is involved. The risks are in marketing and in finance, not in science and technology. This is reflected in the different R&D intensities among industry sectors, and in the fact that a relatively small number of companies account for a large portion of all industrial research.

Two important pressures prevent industry from restricting its goals to only modest technical change. First, the great technically based corporations require large growth opportunities in order to use their capital, their resources, and their organizational structure most effectively and most competitively. DuPont and Monsanto, Exxon and Mobil, IBM and GE must pursue growth that calls for significant technical change precisely because this is the basis for creating major new business opportunities or major new processes.

The second pressure is on companies whose strategic plans are more modest. Even these companies cannot depend on maintaining the status quo. The growth of the technical enterprise worldwide acts to introduce technical change into the markets and processes of these companies, decreasing the option of planning for a slowly changing technological status quo. Biotechnology is forcing change in the chemical and food industries, while developments abroad led to the adoption of new furnaces and mini-mills by the U.S. steel industry. The almost continuous technical advances in microelectronics have sharply reduced the life-cycle of products such as small computers, instruments, control devices, toys, medical equipment. These changes bring great opportunities for newcomers and fast-moving existing firms. And they put pressure on all companies to expand their technical resources in order to absorb and anticipate external technical change.

Thus, the large technically oriented companies, the large companies in more slowly changing industries, and the smaller companies that depend more on adaptation than on creation of technical change—all of these are in the position of planning for business growth that requires increased technical resources. Each category must cope with some fundamental obstacle in achieving the desired technical base, and the obstacle differs for each case. The smaller corporation may not have large enough sales or capital to keep up with all the technical skills which affect their current business and future plans. The large corporation in a slowly changing industry normally has a narrow technical base specific to that industry, may not be able to attract the desired skills from more dynamic technical fields and, for certain process industries, may have serious financial limitations. The 1985 McGraw-Hill survey of industrial research plans shows that R&D expenditures by U.S. industry overall will be 10.9 percent greater than for 1984. The biggest percentage increases will be in the slower changing process industries, steel and nonferrous metals, rising by 25 percent and 29 percent, respectively.[6] This is undoubtedly due to the sharp cutbacks in R&D by metal and mining companies since 1980, and is applied to a much smaller base than, say, electronics. It is a clear indication that the technical needs of the industries were greater than the technical resources remaining after the cutbacks.

Finally, the technically oriented corporation is faced with the nature of technical progress at the frontier. That is, it must constantly increase resources to continue to create significant technical change. But progress in any technical field is being made through the cumulative efforts of the entire technical enterprise. The probability of a single major corporation creating significant technical change in a broad area decreases as the ratio of its technical effort to the overall effort decreases. Thus, even the largest corporation must focus its resources on selected technical niches in which there is reasonable probability of creating significant technical change. The "niche" must be broad enough so that the success of the technical effort will lead to substantial markets or cost savings. This means that the requirements of corporate growth plans exert a steady pressure for increased technical resources. A recent *Fortune* article on the most innovative of the Fortune 500 companies observes that their management "is convinced of the *need* to innovate, regarding new ideas as the essense of long-term survival."[7]

THE DRIVE FOR LINKAGES

Assuming it is true that technological change is necessary for corporate health and vitality and true as well that technical self-sufficiency is no longer possible, what should industry do? The answer would include the following actions.

- Steadily increase spending on industrial research, even in sectors experiencing unsatisfactory growth. (The McGraw-Hill survey referred to earlier shows that real spending for industrial research increased by about 8.3 percent in 1984 and in 1985, compared with an average of about 3.7 percent from 1972 through 1983.)

- Intensify relationships with academe, funding and cooperating in research, thereby gaining access to technical advances and to graduates trained in new skills.

- Increase joint ventures to support major technical developments or to combine the competencies of the different partners.

- Increase collective industrial research involving not just the smaller or low-technology companies, but larger and more R&D intensive corporations.

- Increase the number of acquisitions and mergers motivated by obtaining desired technical strengths.

- Increase licensing.

- Introduce financial mechanisms to facilitate collective research or collective funding of research.

- Promote government policy to facilitate collective industrial research, for example by matching funds and sponsorship, by easing antitrust regulations to accommodate R&D collaboration, or by offering tax incentives for university-industry collaboration.

- Increase internationalization of R&D in all preceding categories.

- Increase the number of industrial research executives named as chief executive officers of corporations.

Many of these practices are already being pursued. The consequences to be expected from changing the conditions of industrial research are many. (See Chapter 9.) Market development, financial opportunities, initiatives by universities, and increased international competition are some of the more important influences determining which of these conditions are most likely to be changed. Linkages with other sectors and other sources within the growing technical enterprise become more desirable and necessary as technical progress continues.

A widely publicized example of linkage was the formation of the Microelectronics and Computer Technology Corporation (MCC) in 1983, proposed initially by William Norris, chairman of Control Data Corporation.[8] Norris was motivated by the perception that Japanese corporations were entering into international competition in microelectronics and computers based on cooperation between the companies and the Japanese government. This presumably put U.S. corporations at some disadvantage, since no single corporation could assemble the resources sufficient to develop technical advantages over "Japan, Inc."

The organization was structured to permit joint sponsorship by all, or some, members of specific technical objectives that would provide a stronger base for some particular product or process direction, but stop short of developing a commercial item. The original members included all the major computer companies except IBM, and a number of semiconductor firms. As of early 1985, there were 21 members. The anticipated level of operation will be approximately $50 million annually.

MCC is not another trade association. It will have its own research facilities and plans to be a major force in strengthening the technologies generic to microelectronics and computers. This may result in patents with significant commercial value, so that their output will not necessarily be in the public domain or flow through traditional public channels for the dissemination of knowledge. The formation of MCC received advance approval from the Anti-Trust Division of the U.S. Department of Justice, though individual projects with a particular group of sponsors may have to be submitted for specific clearance. It is, in fact, a private R&D organization, but focused on the commercial significance of the R&D projects undertaken. It is financed by the member corporations, who receive certain rights and benefits from their sponsorship.

A different form of collective industry action is typified by a consortium of five companies from five countries that plan to develop a new jet engine at a cost of $1 billion.[9] This is a more conventional private action, except that there are five partners, it is completely international, and it is happening more frequently. It incorporates a major technical objective with a business plan as to how the results will be commercialized. The Pratt and Whitney Division of United Technologies Corporation and Rolls-Royce of England will each own 30 percent of the venture. The remaining shares are divided equally among Fiat (Italy), Motoren-und-Turbinen-Union (West Germany), and Japanese Aero Engines.

The participation of United Technologies was cleared in advance by the U.S. Justice Department. Robert Carlson, president of United Technologies has called the venture a trail-blazer in international business. If it is a great success he will be correct, but similar consortiums in the 1970s, to develop systems for ocean mining, were held back by legal as well as market obstacles.

Industry's move toward linkages with other sectors can change the functions of these sectors. The increasing ties with universities are taking on forms today that differ in character from earlier relationships. Multimillion dollar research agreements represent deliberate corporate strategies to facilitate diversification and growth. Some of these agreements are listed in Table 8-1.

Table 8-1. Recent Major Corporate Agreements with Universities.

Company	Institution	Subject	Amount ($ millions)	Years
Hoechst A.G.	Massachusetts General Hospital (Harvard University)	Molecular biology	$50.0	10
Monsanto	Harvard University	Biochemistry	23.0	12
duPont	Harvard University	Genetics	6.0	5
Monsanto	Washington University	Proteins/peptides	23.5	5
Exxon	MIT	Combustion	8.0	10
W.R. Grace	MIT	Microorganisms	8.5	5
Exxon	Cold Spring Harbor Laboratory	Molecular biology	7.5	5

Source: L.S. Peters and H.I. Fusfeld, "University Industry Research Connections," NSB 82-2.

These major agreements have stimulated the university system generally to consider the proper options for deriving commercial benefit from their research output. That alone is a force for change in academia going well beyond simple licensing.

A second growth area for university-industry linkages is the increase in mission-oriented research centers at universities. These permit better meshing with corporate technical needs. The growth of these centers in level of funds has been a primary force moving the major research universities to take on the added function of research management.

Finally, the pressure to seek linkages is reflected in certain government actions. In the United States, we have seen moves to modify antitrust enforcement to encourage R&D cooperation, including legislation to remove triple damage penalties if a cooperative venture is later judged to be in violation. The major program of the National Science Foundation to establish Engineering Research Centers at selected universities is partly motivated to attract and facilitate university-industry linkages. And in Europe, there are national programs (such as the Alvey Program in England) and international programs (ESPRIT in the European Community) which provide matching public funds with private corporations to pursue approved programs in technical areas relevant to microelectronics and to information technology. These are cooperative research programs between industry and government, which bring university resources into the projects wherever reasonable.

These are just a few examples, a few hints of the changing patterns of industrial research and of the pressures behind them. A review of management requirements in a world of rapidly changing technology states:

> Strategists are also coming to realize that two or more heads, and pocketbooks, may be better than one. Many companies now establish joint ventures with other organizations. . . . Yoshi Tsurumi, Professor of International Business at City University of New York, argues that as competition in almost every high-tech business intensifies "any company that limits its R&D resources—money, people, ideas—to its home base will be restricted."[10]

Thus, the status of industry as the largest user-generator of science and technology is a force for growth throughout the technical enterprise. As industry itself pursues linkages with external resources, it stimulates and influences other institutions. Today, the magnitude of industry's R&D expenditures and its initiatives in cooperative research ventures make industrial research the driving force in the process of technical change.

NOTES TO CHAPTER 8

1. Jean-Jacques Salomon, "What Is Technology? The Issue of Its Origins and Definitions," *History and Technology 1*, no. 2 (1984).
2. I refer to industry in general, since isolated instances existed before 1900. Salomon ("What Is Technology") points out that the first known instance where a corporation supported a research program to achieve a desired commercial objective was in 1869. The German chemical firm of BASF (Badische Anilin und Soda Fabrik) began a successful program under A. Von Baer to develop synthetic indigo, first marketed in 1987, 28 years later.
3. "OECD Science and Technology Indicators" (Paris: Organization for Economy Cooperation and Development, 1984).
4. Quoted in the *Wall Street Journal*, May 22, 1985, p. 6.
5. *Dun's Business Month*, December 1983.
6. *Business Week*, June 3, 1985, p. 28.
7. "Eight Big Masters of Innovation," *Fortune*, October 15, 1985.
8. C.S. Haklisch, H.I. Fusfeld, and A. Levenson, "Trends in Collective Industrial Research," Center for Science and Technology, Graduate School of Business, New York University 1984.
9. *Fortune*, November 28, 1983, p. 7.
10. "High-Speed Management for the High-Tech Age," *Fortune*, March 5, 1984.

9 THE NEW CONNECTIONS

It is rare for the growth of a major corporation to occur solely by the expansion of personnel and facilities. Acquisitions, mergers, and joint ventures are traditionally the means to satisfy the needs or the opportunities that a corporation encounters. Impelling these actions is market development. Another company can provide access to a different market, based on geography or on channels of distribution. The second company can have products that add to the first company's line of business. There are many reasons why one plus one equals more than two when different products, different customer base, and different geographic regions are brought together in some form of corporate marriage.

These reasons are not involved directly in the joint venture Corning established with Genentech to explore enzyme technology. They are inadequate to explain why Ford paid $20 million for a minority share of American Robot Corporation or why General Motors invested over $50 million in six companies in robotic vision and one in artificial intelligence.[1] They are only partly involved in the IBM purchase of Rolm. And they clearly have little to do with the major electronic corporations in Europe joining with the European Community to form ESPRIT, the European Strategic Program for Research in Information Technology, or the major semiconductor firms in the United States forming the Semiconductor Research Corporation.

Motivating the flood of such activity throughout industry is the hope of gaining technology not available within the corporation. In the broadest sense, this reaching for external technology is market driven. Either the corporation is faced with the need to meet new market conditions that require major technical advances or, as in the case of Ford and GM, advanced manufacturing techniques are critical for cost competitiveness. There may be opportunities to enter areas of new growth for the corporation which, if not pursued, could result in decline. However, the attention in many recent actions is on the use of outside technology to accomplish corporate strategy.

PRIVATE CHANNELS

The most direct route for obtaining external technology is licensing it from the originator. This adds no new personnel and requires no change in technical direction. Use of licensing is growing, but it has its limitations. A license may be restricted to certain applications or geographic areas, or it may be nonexclusive. For less restrictive licenses the price is higher. In 1985 the Philips Company paid $16 million to Sharp Corporation of Japan to license its technology on liquid crystal displays for manufacture in the Netherlands.[2]

Other means of acquiring technology involve considerations that go beyond technology. Besides the usual forms of access-acquisition, merger, or joint venture-corporations are using venture capital in order to participate in start-up companies that offer new technical developments.

An acquisition or merger calls for organizational, financial, and market considerations. However desirable or important the technology and the technical personnel, a merger or acquisition must make economic sense against far broader criteria. Certainly the purchase of Hughes Aircraft by General Motors for roughly $5 billion was based primarily on corporate diversification plans, giving GM a much larger role in the defense industry. However, the technical base of Hughes was an important consideration since it gives GM "access to advanced Hughes technology that could help it build better cars more cheaply."[3] It has also been referred to as marking "a major new trend, the high-teching of America's old-line manufacturing industries."[4] There is a fundamental difference between licensing technology, where the corporation must identify what it wants, and buying in whole or in part the generator of the technology, where the new personnel can participate in identifying what the corporation can exploit.

A joint venture that is based primarily on technology has fewer associated considerations than an acquisition or merger. Nevertheless, the joint venture almost always involves a business plan for exploiting whatever technical development is to be conducted. Its justification must rest on market and economic criteria.

Advances in areas such as optics and computers, particularly software, often come from small, young companies whose creativity and drive have made them attractive to corporate giants. In 1984, for example, GM bought equity shares in four small vision companies— Automatix (1983 sales: $13 million), Robotic Vision Systems ($2.9 million), View Engineering ($15.5 million), and Diffracto, of Windsor, Ontario ($8 million).[5] It also bought an 11 percent share of Teknowledge, a firm producing software in artificial intelligence,[6] and it has had a joint venture in robotics with Fanuc of Japan.[7]

Just to point out how this process spreads, Intel and Westinghouse then established a joint development program that would produce software to meet the manufacturing automation specifications emerging from integration of the technical advances into GM's production. Intel also set up a program with another California company, Industrial Networking for the joint development of hardware specified by GM for their new systems. Industrial Networking itself is a joint venture formed by GE and Ungermann-Bass, which sells computer networks.[8] Thus, a technology loop is formed by which one very large corporation, GM, reaches out for potentially significant technical advances by a group of vibrant technological entrepreneurs in newly created companies, and this in turn calls for technical support developments that touch upon the resources of Intel, Westinghouse, and GE.

Even before the GM moves, Ford had bought 16 percent of Synthetic Vision Systems for $2 million in 1983.[9] After GM's actions, Ford invested $20 million in "American Robot Corp. stock and that company's technology development."[10] One item of interest in this purchase is the comment by Donald Petersen, chairman and CEO of Ford, that "various divisions within the company have the authority to pursue purchases of technology outside of Ford." The American Robot investment came from the Ford's Diversified Products division, which manufactures electronic components for the automobile divisions. Petersen added, "If the best way to get technology is through acquisitions, we have an open door policy."

The automobile companies represent the other side of the coin in the development of technical linkages. The small, creative generators of new technical advances gain the resources necessary to broaden their base and develop applications. Importantly, working relationships are established between the basically R&D companies and the large users which provide the feedback necessary to guide a new technology to a solid foundation.

This is not an American phenomenon by any means. The Daimler company in Germany, moving to diversify into high-technology industries, is pursuing acquisitions to add the desired technical base. In February 1985 it bought control of the large German manufacturer of big engines, MTU (Motoren-und-Turbinen Union) for $225 million. This is the company which is part of a five-company, five-country consortium to develop a new jet engine. It is purchasing a controlling interest in Dornier, which is active in space programs and has indicated its desire to work with the U.S. Strategic Defense Initiative Program. Dornier's research in electronics and materials "will help Daimler develop advanced technology for cars, trucks and buses." The broader significance of these major purchases "is a sign that Eurosclerosis may be ending. Top European companies are now aware of their deficiencies in high tech and are pushing to fill the gaps. If more companies follow in Daimler's footsteps, the result could be a newly competitive Europe, something that seemed impossible just a few years ago."[11]

A large merger is rarely based on technology alone, and access to external technology is not usually accomplished by mergers, particularly when large corporations are involved. There are far too many other considerations of structure and personnel. Much more common is the joint venture. Two or more parties can agree to pursue some technical development and an associated business plan without serious interference with existing operations.

One important basis for a joint venture having a technological emphasis is the need for the generator of a new technical advance in one industry to work closely with the user of that advance in another industry. IBM has entered into several joint ventures to develop applications of computers in factory automation systems.[12] It has set up joint ventures with Cincinnati Milacron and other U.S. manufacturers of factory equipment. It also has an agreement with the Italian firm STET-SFT, a major supplier of telecommunications and factory equipment, to develop software and hardware for factory automation.

Another great technically based company making extensive use of this type of joint venture is Corning Glass. Corning has had great success with Owens Corning Fiberglas set up 50 years ago to combine Corning's technology with Owens-Illinois's experience in applications. Dow Corning was another successful joint venture, combining Corning's advances in glass technology with Dow's competence in chemical processes and marketing.[13]

Today, with little sales growth and declining profits, Corning is seeking a broader technical base to permit growth in new areas. Its optical waveguides are the basis for growth. A joint venture, Siecor, was set up in 1979 with Siemens of Germany, in order to develop complete systems with a major company from the communications field.

Corning's glass research established that a column of glass beads could be used to immobilize enzymes, so that they can be removed from a solution and reused. This has areas of application in food processing. In order to pursue this, a joint venture with the Kroger Company was set up to develop such applications. The joint venture, Nutrisearch, has developed processes for producing protein-rich sweeteners, and syrup that is convertible to yeast, starting from a waste product of cheese processing.

Given these experiences in enzyme technology and food processing, Corning then set up a joint venture with Genentech in biotechnology. This company, Genencor, will develop processes for using recombinant DNA to produce enzymes which could then be used in industrial processes. This would give Corning a solid base in genetic engineering.

The joint venture can be a very practical mechanism to link technology internationally. It avoids the more complex problems of acquisition, although buying a minority share of equity can be a compromise approach. For example, RCA and the Sharp Corporation of Japan have a $200 million joint venture which will result in a U.S. plant to produce products using C-MOS technology based on existing competencies and future developments.[14] A deliberate program to gain access to new advances in telecommunications has been initiated by the Spanish telephone company, Telefonica "in an effort to update its technology."[15] It has established technical exchange agreements with AT&T, Corning Glass, Fujitsu, and L.M. Ericsson of Sweden.

An interesting loop of joint ventures is being created by AT&T and Olivetti. AT&T purchased 25 percent of Olivetti for $260 million in 1983. This opens new markets for AT&T equipment and provides Olivetti with access to AT&T technical advances. Olivetti later announced that AT&T will pay an additional $250 million for a new Olivetti product development.[16]

Both Olivetti and AT&T are developing other linkages that combine technical developments, technology flow, and marketing across national boundaries. Toshiba Corporation bought 20 percent of Olivetti of Japan, which will now have a link to Toshiba technology. It will obviously broaden Olivetti's market in Japan and provide outlets for Toshiba technology in Europe. With reference to the AT&T equity in Olivetti, Vittorio Levi, managing director of Olivetti, said, "The two agreements are mutually compatible and may even make it feasible for Toshiba, Olivetti and AT&T to collaborate with one another in the Japanese market in the future."[17]

Olivetti now has investments in over 30 high-technology companies. A number of these are in California. In February 1985 it bought control of Acorn, a microcomputer company in England. An agreement was made with Xerox in April 1985 for distribution of Olivetti products by Xerox.

AT&T has been seeking alliances and linkages as it pursues markets and breaks out of the domestic limits it observed prior to divestiture of the local phone companies. The largest joint venture it has entered into is that with the Philips Company in the Netherlands.

One interesting sidelight of history is that Nippon Electric Corporation (NEC), a major producer of semiconductors and telecommunication equipment, was begun as an international joint venture. It was set up in 1899 by the Western Electric Company and two Japanese entrepreneurs. The Western Electric shares were sold in 1925 to ITT, which gradually disposed of them.[18]

European corporations are increasingly active in joint ventures. They are often driven by both the need to link up with an external source of technology and the need to reduce very considerable costs of achieving high-technology advances. Marketing can be an important consideration as part of a business plan for producing some financial return from the venture.

Thus, Philips and Siemens, both giants in electronics, have established a joint research program to pursue long-term technical efforts generic to semiconductors. This is part of a program between the Dutch and German governments. Siemens, International Computers Ltd. (ICL), and Bull of France have jointly set up a research laboratory in Munich, the European Computer-Industry Research Center, to conduct research in artificial intelligence.[19] These efforts are devoted solely to R&D. However, the common ownership of commercially useful results plus the relationships built up among partners could lead easily into business relationships for exploiting technical advances.

The reasoning behind these European linkages among large high-technology companies, and particularly with respect to their participation in ESPRIT is perhaps best summed up in the form of questions raised by the Philips representative for ESPRIT.

Do any of the European companies by themselves have the power to become a major factor in the open world market?

Can any country's academic and industrial research marshal the necessary scientific and technical competence, getting at the necessary level in the short time frame that is required?

Can local industry make the necessary investments to set up and sustain product development, manufacturing and commercial organization, to reap the benefits of R&D on a world scale? Such investments are orders of magnitude larger than any money devoted to advanced R&D.[20]

These questions sum up the pressures behind recent actions in microelectronics and computers, not only in Europe, but also in the United States and Japan. One answer to these questions comes from the chairman of Philips, Wisse Dekker, who remarked: "We need the best global technology we can get if we're going to stay competitive."[21]

Especially significant international linkages are those between American and Japanese firms. It is of particular interest because of the rising concern about competition in international trade between the two countries. The general public and members of Congress are probably not aware of the extensive flow of technology between the United States and Japan.

An initial study carried out at the NYU Center for Science and Technology Policy focused on the extent of U.S.-Japan technical exchanges in four industries: electronics, pharmaceuticals, chemicals, and metals. The objectives and significance of these exchanges were discussed with a number of the U.S. companies involved.[22] The study looked at agreements that embodied some transfer of technology, not a purely distribution agreement.

For examples of technical agreements in the electronics industry, see Table 9-1. The agreements reported in the business press increased from two in 1979 to forty in 1984.

Technical exchanges in pharmaceuticals and chemicals show similar growth. There is a flow of technology in both directions, though the balance of flow is different for different periods and for different industries. The companies involved represent a wide diversity of strengths. There are small and large companies, leaders and followers. A number of the Japanese companies are trading companies, which serve as middlemen to locate the appropriate Japanese company to provide or absorb the technology covered by the agreement.

The important point in all this is that Japan and the United States have established mechanisms for access to roughly the same common base of science and technology. Obviously, the highly sophisticated companies may not place their very latest advance into a technical exchange And the "common base of technology" must be interpreted statistically, not individually. That is, one or a few U.S. companies may have access to a particular piece of Japanese technology, and vice versa. But collectively, the United States has access to Japanese advances through these exchanges, with whatever restrictions a business agreement may contain, just as Japan has access to U.S. technology through the same mechanism.

I do not believe these interactions, perhaps even interdependencies, are well understood or much thought about. They are certainly lost in the intense unhappiness with trade relations between the two countries. And they are greatly complicated by the growing investments from one side in the other side, so that manufacturing plants are operated in one country but owned by a company from the other country or, increasingly, by a partnership.

An apocryphal story is told by the commercial counselor from Japan's Washington embassy. According to the story, an American walks into a local appliance store and says: "I want to buy a TV set, but I want one from a domestic manufacturer." The owner replies: "O.K., would you prefer a Sony made in Arizona or an RCA made in Mexico?"

Table 9-1. Technical Agreements between U.S. and Japanese Electronic Companies, 1979–1984.

U.S. Firm	Japanese Partner	Direction	Details
		1979	
Energy Conversion Devices and Burroughs	Sharp	U.S.–Japan	Nonexclusive worldwide licenses to use semiconductor "ovonic" memory devices.
Siliconix	Teijin Advanced Products	U.S.–Japan	JV (Nippon Siliconix): Marketing semiconductor products
		1980	
Intel	NEC	Japan–U.S.	Floppy disk control
RCA	Toshiba	Japan–U.S.	1K/4K DRAMS for CMO5 SOS
Motorola	Aizu–Toko	Joint ventures	NMOS/CMOS technology
		1981	
Intel	Sanyo	U.S.–Japan	Licensing contract to manufacture 8086 and 8088 16-bit CPUs and 8-bit CPU
Fairchild	Sanyo	U.S.–Japan	Licensing contract to manufacture 4-bit CPU
Westinghouse	Mitsubishi	Japan–U.S.	Negotiating JV for producing 64K RAMs, eventually Mitsubishi builds own plant (1983)
Intel	Fujitsu	U.S.–Japan	Technology exchange. Fujitsu will manufacture Intel's IAPY 86, IAPX 88 and second-source 8086, 8088, 8089 microprocessor worldwide nonexclusive license for manufacturing and marketing.

(*Table 9–1. continued overleaf*)

Table 9-1. continued

U.S. Firm	Japanese Partner	Direction	Details
Rockwell	Sharp	Japan–U.S.	CMOS process
Motorola	Hitachi	Japan–U.S.	Hi-CMOS process
SMC	Hitachi	U.S.–Japan	Complamos process for NMOS
Intel	Mitsubishi	U.S.–Japan	8086
LSI Logic	Toshiba	Joint ventures	CMOS gate arrays and wafers
TRE Semi-	Tokyo Electron	Unknown	Steppers, lithography
Thermco Prod.	Tokyo Electron	Unknown	Oxidation, CVD, diffusion
Morton Chemical	Tokyo Ink Mfg.	Unknown	Epoxymoldings compound
CTI-Cryogenics	Ulvac	Unknown	Cryogenic pumps, sputterers
Gen Rad	Tokyo Electron	Unknown	IC test systems
1982			
Standard Microsystems	Fujitsu	U.S.–Japan	Worldwide nonexclusive cross-licensing of patented SC technology
Zilog	Toshiba	U.S.–Japan	Technology exchange and development, second-sourcing Toshiba second-source of Z80 (8 bit), Z800 (16-bit) MPs. Zilog gains advanced OMOS versions of Z series and OMOS fabrication process technology (64K RAM) equivalent.
Trilogy	Sumitomo	U.S.–Japan	Joint venture, marketing subsidiary

			IC technology
LSI Logic	Toshiba	U.S.-Japan	
Hewlett-Packard	Hitachi	Japan-U.S.	First 64K DRAM production agreement. Hitachi supplied masks, production equipment and some engineers, H-P will produce for own use only.
Rockwell	Ricoh	U.S.-Japan	Technology exchange agreement. Ricoh provides 32- and 64K OMOS EPROM technology in exchange for 8-bit microprocessor technology.
VLSI Technology	Ricoh	U.S.-Japan	Technology exchange agreement. Ricoh provides ROM masks for 64-256K OMOS technology, in exchange for ROM 64-256K NMOS tech.
Rockwell	Sharp	Japan-U.S.	Sharp supplies IC tech.
Fairchild	Hitachi	Reciprocal	F6856 synch-protocol Hitachi's 6845 CRT controller 64 K DRAM
Intel	NEC	U.S.-Japan	8022/8253A/8259A and MuPD765 MuPD7201/MuPD7220
Zilog	Toshiba	Reciprocal	16K CMOS SRAM and Z80/Z800/Z8000
Ungermann-Bass	Fujitsu	Joint venture	Ethernet chips
AMI	Asahi-Chemical	Joint venture	CMOS gate arrays
Sperry	Mitsubishi	Joint venture	64K DRAMS, gate arrays
Micro-circuit Engineering	Kanematsu Semiconductor	Joint venture	Gate arrays

(Table 9-1. continued overleaf)

Table 9-1. continued

U.S. Firm	Japanese Partner	Direction	Details
Westinghouse	Mitsubishi Electric	Unknown	ICs for factory automation
General Signal	Omron Tateishi	Unknown	Stepper, probes, dicing
Materials Res.	Midoriya Electric	Unknown	Sputterers for 256K DRAMS
Varian	Tokyo Electron	Unknown	Ion implanters
GCA Corporation	Sumitomo Shoji	Unknown	Stepper and wafer-tracks
Aehr Test System	Various firms	Unknown	IC burn-in systems
Contact System	Tabai Works	Unknown	Automatic insertion
Airco Temescal	Tokyo Ohka	Unknown	Plasma etchers
Varian	Analytica	Unknown	Analytical instruments
Perkin-Elmer	Kanematsu Semi.	Unknown	E-beam lithography, aligner and metal sputterers
Smiel (Synamit Nobel)	Japan Silicon	Unknown	Gettering technology
Veeco Inst.	Kokusai Electric	Unknown	Leak detection, ion million, ion implantation
_1983_			
National Semiconductor	Oki	Japan–U.S.	JV, National to produce, market, and supply to Oki's 64K RAMS. Joint development of 64 and 256 RAM and MOS products. Oki only one of "bix six" Japanese without own SC U.S. Plant.

Standard Microsystems	NEC	Japan–U.S.	Worldwide nonexclusive cross-licensing, Standard is 2nd source for 4 of NEC's microprocessor related MOS VLSI IC products; UPD service controllers.
American Microsystems	NEC	Japan–U.S.	Second-source to produce NEC's 4, 8, 16-bit microprocessors. AMI will supply telecommunications LSI information.
Intel	NEC	Japan–U.S.	Second-sourcing NEC tech.
American Microsystems Inc. (Gould)	Asahi Chemical Industry	U.S.–Japan	JV in Japan and U.S. (Asahi Microsystems, Gould 51%) for research, design, development, manufacture, sale of custom-made MOS chips. Technology from AM, capital from Asahi.
Standard Microsystems	Toshiba	U.S.–Japan	Worldwide nonexclusive cross-licensing. Toshiba pays SM.
Zilog	NEC	Japan–U.S.	Technology exchange following patent dispute.
Motorola	Hitachi	Japan–U.S.	CAD tech. for MPUs
RCA	Nippon Denso	U.S.–Japan	CAD ICs process technology
Tektronix	NEC	Japan–U.S.	B-bit MPUs (NMOS/CMOS)
Zilog	Sharp/Toshiba	U.S.–Japan	Z8000 family
Intel	Sanyo	U.S.–Japan	8086/8051
VLSI Technology	Ricoh	U.S.–Japan	64K/128K/256K mas ROMS
Applied	Sankyo Seiki Mfg.	Unknown	Material-handling robots

(*Table 9-1. continued overleaf*)

Table 9-1. continued

U.S. Firm	Japanese Partner	Direction	Details
Eaton	Sumitomo Heavy	Unknown	Ion implantation
Lam Research	Tokyo Electron	Unknown	Plasma etchers
Ungermann–Bass	Mitsui Shipbuilding	Unknown	CAD/CAM development
Machine Tech.	Nippon Tairan	Unknown	Manufacturing equipment
Union Carbide	Mitsubishi Metal	Unknown	Chemical compounds, oxide single crystals
Centes	C. Itoh	Unknown	Manufacturing equipment
Venture Business	C. Itoh	Unknown	CVD for silicide production
National Semiconductor	OKI Electric	Joint venture	64K DRAMS
Stanford Applied Eng.	Mitsubishi Chemical & Kyoshin kogyo	Joint venture	IC Sockett, connectors
SMOS Systems	Suwa Seikosha	Joint venture	CMOS gate arrays
		1984	
Texas Inst.	Fujitsu	Reciprocal	Mutual second-sourcing of CMOS and bipolar gate arrays.
Zilog	Toshiba	Japan–U.S.	Toshiba supplying CMOS 8-bit MPU on OEM basis
Motorola	Toshiba	U.S.–Japan	Second-sourcing of Motorola 16-bit MPU
Intel	Fujitsu	U.S.–Japan	Fujitsu takes partial charge of 16/32-bit MPU development
Intel	Tokyo Sanyo	U.S.–Japan	Sanyo captive use of 8086/8088/8051

Zilog	NEC	U.S.–Japan	Settlement of Z80 patent infringement
Intel	OKI Electric	U.S.–Japan	Oko captive use of Intel MPUs, MCUs, and MPR LSIs
Zilog	NEC	Reciprocal	Mutual sourcing of NEC V Series and Zilog Z8000 32-bit MPU
AMI	Hitachi	Japan–U.S.	Second sourcing of codecs
Standard Microsystems	OKI	Reciprocal	Cross-licensing of all patents and patent applications
Inmos USA	NMB Semiconductor (Minebea)	Unknown	CMOS 256K DRAM
Monolithic Memories	Fujitsu	Japan–U.S.	Second sourcing of Fujitsu's TTL gate arrays
Rockwell	Ricoh	Japan–U.S.	Ricoh supplying CMOS 64K EPROMS to Rockwell
Intel	Fujitsu	U.S.–Japan	Second sourcing of EPROMS 8-bit, single-chip MCUs, and MPUs
Panatec R&D Corp. and Modular Semiconductor	Ricoh	U.S.–Japan	Joint development, technical exchange, and mutual sale agreement for 256K DRAMS
Intel	Fujitsu	U.S.–Japan	Second sourcing of Intel 16-bit MPUs and peripherals
Monsanto	Sony	Japan–U.S.	Magnetic field Czochrals method
Custom MOS Arrays	Ricoh	Joint venture	CMOS gate array design and wafer production

(*Table 9–1. continued overleaf*)

Table 9-1. continued

U.S. Firm	Japanese Partner	Direction	Details
Tektronix	NEC	Joint venture	Tektronix marketing of V Series support MPU systems
Microtec Res. Corporation	Zax Corporation (start-up)	Joint venture	Joint development of System Z MPU support systems
Motorola	Toko	Joint venture	Read/write amplifiers for flexible disk drives
Panatec R&D Corporation (US)	Ricoh	Joint venture	Joint development and marketing of ICs
Digital Res.	NEC	Joint venture	Joint marketing of CP/M operating system for V Series
Corvus Systems	NEC	Joint venture	Joint development of LSI for CMOS single-chip controller
RCA	Sharp	Joint venture	Design center and wafer fab on West Coast of U.S. RCA owns 51%
Kayex USA	Koyo Lindberg	Unknown	Crystal growing, slicing, polishing equipment
Perkin-Elmer	Kanematsu Semi.	Unknown	Dry etchers, steppers
Anicon	Sumitomo	Unknown	CDV equipment
Hemlock Semiconductor (Dow Corning)	Shinetsu Handotai	Unknown	Polysilicon
Hewlett-Packard	Yokogawa-Hokushin	Unknown	Liner IC test equipment

Veeco	Kokusai Electric	Unknown	Ion beam etchers
Varian	Gartec	Unknown	Sputterers
Genus USA	C. Itoh	Unknown	CVD equipment
March Instruments	Samco International	Unknown	Ethcers, small equipment
Integrated	Kishimoto Trading	Unknown	Wafer transport systems
Diamond Image	Kinematsu-Gosho	Unknown	Exposure equipment
BTU Corp.	Ulvac Corporation	Unknown	CVD, diffusion, oxidation
Pacific Western	Kanemitsu	Unknown	CVD equipment
Daisy Systems	Seiko Instruments & Electronics	Unknown	CAD work stations
Stanford	Plasma Systems Tazmo Co. Musashi Mfg.	Unknown	Plasma etchers, other

Source: L.S. Peters and H.I. Fusfeld, "Technical Exchanges Between U.S. and Japanese Industry," Center for Science and Technology Policy, Graduate School of Business Administration, New York University, 1986.

Technology exchanges between U.S. and Japanese companies should be a critical element of government policy, which normally focuses on trade alone without examining the technical base. There is great economic significance in the fact that the two countries that are strongest in both generating and using science and technology are reinforcing each other's technical strengths. These linkages strengthen the entire technical enterprise and can be an important factor in continuing to foster technical advances.

PUBLIC CHANNELS

The actions of government, domestically and internationally, focus more on funding and facilitating R&D rather than on conducting R&D within government laboratories. Nevertheless, both of these functions play a role in expanding linkages within the technical enterprise.

Government Research Laboratories

Government laboratories are not involved in very many formal or active types of technical linkages. Laboratories committed to broad programs of basic research maintain contact with their fields and their colleagues through traditional professional channels. These would include such major institutions as the National Bureau of Standards and the National Institutes of Health in the United States, and the Centre Nationale de la Recherche Scientifique (CNRS) in France.

International cooperation links similar laboratories from different governments into common programs. Major research facilities become the basis for joint research programs needed to pursue costly research. This is the case of Centre Européene pour Recherche Nucléaire (CERN) near Geneva, which is a joint venture for the countries participating.

These are a modest number of interactions with the private sector on the part of several government laboratories that have a specific mission. The Bureau of Mines has a long history of relations with companies in metals and mining, performing cooperative studies of mine slope stability, for example. In a sense NASA is a major creator of linkages among all the contractors involved in space flight, with all technological contributions coming together at a launch.

The Bureau of Standards represents a broad-based laboratory with special facilities and competences. It is itself a nucleus for linkages with comparable organizations in many other countries through cooperative programs and scientist exchanges. It has created ties with industry through its Industrial Research Associates Program.

There has been a reluctance on the part of government laboratories to develop cooperative research programs with individual companies. This is due to the sensitivity to offering possible competitive advantage to one company with taxes paid by its competitors. There are, however, a number of examples, though still not very many, where cooperation has been established with industry associations, more acceptable politically. Much of this takes the form of programs at the government laboratories which are funded by the associations. For example, 36 contracts for various types of technical activities are funded at laboratories of the Department of Energy by four associations. These include the Electric Power Research Institute (EPRI) and the Gas Research Institute (GRI).[23]

Funding and Facilitating R&D

Funding R&D and facilitating it is the principal role of government in the use of public channels to promote or support linkages. Facilitation can take many forms, most of which require money. The government can provide both the legal umbrella and the funds. It can also offer neutrality in bringing together parties in a technical development with their own vested interests.

In one sense, government provides some of the glue that ties the technical enterprise together. The patent system is a critical element in the infrastructure of science and technology, since it provides a major part of the currency used in technical exchanges. The existence of international patent agreements is one of the principal reasons why technology can flow "freely" throughout the worldwide system.

Even more critical is the role of government in assembling a very broad information base about technical programs and outputs which is the core of the communications within the technical community. There is first the data base of government R&D, much of which is published and disseminated by the National Technology Information Services (NTIS) of the Department of Commerce. There have been, and still are, government subsidies for translations of foreign journals. Probably the largest such activity in the world is the USSR for absorption within the Soviet technical system. There are programs by NSF and others to develop computer-based networks to disseminate technical data.

Indirect government subsidies support many professional activities for diffusion of science and technology. For example, the dominant government support for university research normally includes funding, at least in the United States, for expenses necessary to attend technical meetings, and publication charges for professional journals. These items make up the infrastructure of our technical enterprise, and their continued contributions and growth are facilitated by indirect government subsidy.

Direct linkages that focus more technical resources on problems of broad public interest do result directly from government actions initiating or sanctioning such actions. This is particularly evident internationally, where technical cooperation among countries to achieve national objectives of each is perhaps a unique function of government.

An excellent example of government effectiveness to facilitate the creation of these linkages is the International Energy Agency (IEA). Established by governments, this organization serves to coordinate specific projects involving two or more of its member countries. Projects are carried out under separate agreements, with IEA acting as an umbrella for a range of activities. They include development of superconducting magnets for fusion power, high-energy physics, solar power systems, wind-energy conversion, and development of geothermal resources. It has established a technical information service for biomass conversion. IEA arranges for exchange of researchers and has established a direct communications network among researchers. These are all activities in which government facilitation can be highly productive. It results in a network that brings together resources and expedites technical progress. Governments participating in the IEA are the United States, the Federal Republic of Germany, Australia, Belgium, Canada, Italy, New Zealand, Spain, the

United Kingdom, Switzerland, Sweden, Denmark, Ireland, Mexico, the Netherlands, and Japan.[24]

There are many such international technical linkages created by governments to encourage cooperation and diffusion of science and technology in particular fields. They are not often as broad-based or as active as IEA, but they have the same general pattern of focusing international efforts on problems of broad public concern.

More directly relevant to the issues discussed in this book is the increase of government programs, both national and international, that create linkages among corporations in order to strengthen some national or regional industries. These programs arise in response to the same type of pressures which are causing industry to reach for external technology through its private actions.

The European Communities (EC) has been perhaps the most active organization on the international scene in this regard. It is an instrument of the national governments which formed it from the three earlier groups, namely the European Coal and Steel Community, European Economic Community, and Euratom. It has substantial funds contributed by the member countries, roughly $22 billion in the 1983 budget.[25]

The EC itself represents a form of technical linkage through its internal research, largely in energy, and by facilitating technical cooperation in different forms. One traditional mechanism for technical cooperation was the establishment in 1970 of its program called COST, for Cooperation in the Field of Scientific and Technical Research. It has the broad objective of strengthening technical development of the member countries by pooling resources where feasible. The programs eventually agreed on included telecommunications, materials for turbines, and pollution control. These were of public interest and generic to a mission area, not product oriented.

Recently, the EC has initiated less traditional forms of technical cooperation that (1) aim at applying increased technical resources to economic growth, and (2) involve corporations more directly, as performers and recipients of the technical programs. Probably the largest and most significant example to date is the establishment in 1984 of European Strategic Program for Research in Information Technology (ESPRIT). The objectives and implications of ESPRIT bear directly on the principal themes of this book.[26]

The motivation for the program came from two lines of thought:

1. Corporations in the United States and Japan were far ahead in semiconductor and computer technology, reflected in their exports to Europe.

2. The technical resources of any individual corporation headquartered in Europe were inadequate to narrow this "technology gap."

According to the general background statement, "in order to reverse the trend of increasing reliance on importing technology . . . joint strategic long-term research planning and the concentration of resources [can help] by

    (i)   ensuring that research teams achieve the critical mass to obtain results;
    (ii)   enabling optimization of resources that will result in reducing duplication and widening the spectrum of research tackled;
    (iii) reducing the timelag effect caused by reliance on imported technology;
    (iv) paving the way to the definition and adoption of standards of European origin.[27]

One of the many interesting aspects of ESPRIT is that the cooperative effort was pushed by industry, though strongly supported by Etienne Davignon, vice-president of the Commission of the European Community. It began with the 12 leading European corporations in their field:

United Kingdom:	General Electric Company
	International Computers Ltd. (ICL)
	Plessey Company
France:	Compagnie Générale de l'Électricité (CGE)
	Cie, des Machines Bull
	Thomson—CSF
Federal Republic of Germany:	AEG—Telefunken
	Nixdorf Computer
	Siemens
Netherlands:	N.V. Philips Gloeilampenfabriken
Italy:	Olivetti
	Societe Torinese Esercizo Telefonici (STET)

ESPRIT is thus an approach to obtain concerted effort among these major technology-intensive companies, possessing some of the largest research organizations in Europe, plus matching funds provided by the EC for approved projects, in order to permit technical progress in Europe that can make a significant difference in the worldwide technical enterprise. In the five-year period beginning in 1984, the total program will spend 1.5 billion ECUs, roughly $1.25 billion, 50 percent provided by the companies and 50 percent by the EC. (The ECU, the monetary unit of the European Community, is an index made up of a weighted average of the principal currencies within the EC. In 1984, one ECU was worth about $0.84.)

This amount is significant. On the other hand, the total five-year program represents about 15 months of R&D at Philips alone, or about a half year at IBM. To be meaningful in achieving the objectives of ESPRIT, there must be more to it than this substantial but relatively small increment to total R&D expenditures. There is, or seems to be, even though the program is too new to permit any real evaluation.

The potential for significant impact of ESPRIT appears to derive from these separate aspects:

1. By focusing on five areas, then selecting specific projects in each, the funding can be quite significant for those technical niches.

2. Positive technical results from ESPRIT, which will bring into the projects universities and small high-technology firms, can stimulate further development work within European research organizations, thus adding leverage to the ESPRIT funds.

3. Development of European standards for components and equipment arising from the ESPRIT program can be a critical factor in creating a Europe-wide market with distinct competitive advantages for European firms.

4. Perhaps most important for the long-term, ESPRIT encourages clusters of companies to cooperate on particular projects, and thereby provide a proprietary interest and working relationship that can easily lead to joint business ventures for the exploitation of those technologies.

The sum of all this is that ESPRIT can lead to the creation of much larger "user-generators" of science and technology, with a resulting acceleration both of technology and of applications. The actual projects funded by ESPRIT are described as "precompetitive," since they are in the stages prior to the design and engineering of specific products. Nevertheless, they can be the basis not only for more concentration by the companies on the competitive technical stages, but for joint efforts by two or more companies which can combine marketing and manufacturing capabilities in the selected areas.

The success in at least launching ESPRIT has emboldened the EC to push on other fronts to strengthen European technical resources by creating linkages among relevant organizations. A similar program is being considered in biotechnology, though the newness of that field commercially makes it difficult to identify viable research programs. Initial emphasis may therefore rest on data collection and dissemination, conferences and general strengthening of communication in the field.

The EC is moving ahead with Basic Research in Industrial Technology for Europe (BRITE). This will allocate $166 million over four years to companies and research institutes to develop new industrial processes.[28] It is a more generic program than ESPRIT, and spreads less money over more industries. Again, however, it is concentrating funds and attention in a way that can create leverage. By working directly with companies, it establishes the feedback necessary to improve effectiveness of R&D, and may provide competitive advances in selected processes.

Finally, there is the French advocacy of a European-wide research program called Eureka. Politically, this has been advanced as an alternative to the U.S. proposal that allied countries in Western Europe join in some of the technical projects planned under the Strategic Defense Initiative (SDI or "Star Wars"). The political arguments aside, the reasoning behind Eureka is of direct interest to the contents of this chapter. It goes something like this: The technical efforts to be pursued under SDI will strengthen U.S. corporations in many advanced areas such as lasers, microelectronics and computers, thereby increasing the technical lead of the United States over Europe. European companies feel some pressure to join with the United States in order to gain access to this new technology, but are uncertain about the conditions of transfer and use. Since the European interest is with increasing the technical base for economic growth, perhaps a more efficient use of its technical resources would be to develop cooperative programs in areas targeted to preferred product and process opportunities.

The reasoning echoes the themes of declining company self-sufficiency together with the pressure to increase resources. The implementation of Eureka is unclear, since a simple program of cooperative R&D does not automatically result in effective integration of results within the industrial system. The proposal itself is of interest simply to highlight the explicit acceptance of the themes discussed in this book. The author of a *Wall Street Journal*[29] article on the background for Eureka commented:

> In Europe . . . government ministers ponder ways to bolster cooperative research and close Europe's technology gap with America and Japan. . . . "There's a tremendous emphasis on this (coordinating research efforts) by heads of government," says Geoffrey Pattie, Britain's minister for information technology. "It seems to come up in nearly all the summit conferences."

> The underlying logic of this cooperation is simple: high-tech research is too expensive for any one company or country in Europe to handle alone, so it has to be done jointly. And the interest in fostering research increases as more politicians see the high-tech industry as a way out of double-digit unemployment rates.

Facilitating the concentration of technical resources is not left to international bodies alone. National governments are deeply concerned about the ability of their domestic corporations to strengthen their technical base in growth areas, and thus provide improved employment opportunities.

The Federal Republic of Germany is pursuing a five-year, $1.2 billion plan to strengthen microelectronics, advanced computers, and communications technologies. Major electronics companies in West Germany will add their own funds to set up cooperative relations with the government program, and expedite conversion to commercial use.[30] This effort is in addition to West German participation in ESPRIT, and the private agreement of Siemens, Bull (France), and ICL (England) to establish a basic research laboratory in Munich.

Sweden is focusing on its ability to pursue industrial application of custom-designed integrated circuits. Funding of 714 million kroner (roughly $90 million) will be provided, 75 percent by the government and 25 percent by industry. Two companies expected to be most involved with the program are ASEA-HAFO and AB Rifa, a wholly owned subsidiary of L.M. Ericsson. These are the largest makers of custom-designed integrated circuits in Sweden. Government funding will come from the Board for Technical Development (STU). The program is similar to programs in France, the United Kingdom, West Germany, and Japan, but is only a small fraction of their size.[31]

One of the major national programs in microelectronics is that of the United Kingdom. It is the Alvey Program, named for John Alvey, who chaired a commission to consider appropriate national action in that field. It is a five-year program begun in 1983 with about $500 million funding. Of this, 50 percent is government money, 50 percent industry. University participation will be government-funded. It emphasizes four areas: very large scale integration (VLSI), software engineering, intelligent knowledge-based systems, and man-machine interfaces.[32]

The Alvey Program establishes a consortium for each technical effort. A consortium might include, for example, one or more companies, a government laboratory, and a university. The Program consists of the network joining these consortiums together. It is a mechanism concentrating technical resources on complex problems. Its intent is to correct a problem in the British technical system described by the Alvey Commission: "Compared with our competitors, our [the British] overall effort is badly fragmented. The interface between industry and the research community is nowhere near as productive as in the U.S. for example. And our industry does not collaborate on basic research to the same extent as Japan."[33]

As a last example of national programs, consider Japan.[34] The principal feature is the existence of groups of research associations of several types, which provide support for, and linkages with, different industry sectors. These have come into being over a period of years as part of a general government policy to strengthen the national technical base, with particular attention to the needs of different industry sectors. These have been summarized by Roy Rothwell, a colleague on the staff of the Science Policy Research Unit (SPRU), University of Sussex, as follows:

Category 1:   18 government centers for industrial technology. These perform applied R&D and are attached to the major technical ministries. Informal ties to relevant companies are well developed, and some cooperative programs with industry have resulted from research at the centers.

Category 2:   600 local centers attached to municipal or provincial government. They perform testing, R&D, and training for small local firms. The companies pay for some of the technical services, amounting to less than 7 percent of the cost of any one center.

Category 3:   Semipublic centers which are industry specific. These have close relations with the companies affected. Technical and information sources are provided. Research programs are established in collaboration with groups of companies. Company personnel are active on management and technical committees. Some financial support comes from industry.[35]

Perhaps the best example of the use of consensus as a form of making progress is in the formation of a different type of organization—the research association. This has been the subject of a study by the Hudson Institute,[36] and is the most relevant to the newer forms of linkages discussed in this chapter. These organizations are formed by companies with an interest in some common research area. The initiative for establishing a research association can come from the government, which then solicits research proposals and interest from appropriate companies, or from the companies, which can recommend to the government that it take the necessary actions. Research is conducted normally at company facilities.

This mechanism is carried out under the Industrial Technology Association Law. Of the 54 associations initiated, 38 are operative today. Under this legislation, which includes authorization by Ministry for International Trade and Industry (MITI), government funds can be made available to these associations either in whole or in part or as a loan which may be repaid when a project leads to commercial success.

An example of industry initiative is the Biotechnology Association. Five companies held a Biotechnology Round Table in November 1980 and concluded that a research association was desirable. With the addition of other firms, they stated the need for this in July 1981. In August 1981 MITI officially requested proposals for biotechnology research, and 14 companies responded as an incipient Biotechnology Association. The group was officially registered by MITI in September 1981 awarded a research contract and assigned

Table 9-2. Selected Research Associations in Microelectronics.

Association	Established	Member Companies
VLSI Development Association	1976	7
Medical and Welfare Equipment Development Association	1976	35
Engineering Research Association of Flexible Manufacturing System Complex with Laser	1978	18
Basic Technology Research Association of New Computers	1980	10
Research Association of Laser Applied Measurement and Control System	1981	8
Research Association of High Speed Computer System	1981	5
Research Association of New Functional Devices	1981	10
Research Association of New Computer System	1982	8

Source: Data received from M. Uenohara, Executive Vice President and Board Member, Nippon Electric Company (NEC).

responsibility for biotechnology within a broader New Materials Project.

The area of microelectronics has been a major emphasis for research associations in Japan. In recent years, a number of groups have appeared which are directly or indirectly involved with microelectronics. Eight of the most relevant are shown in Table 9-2.

The oldest and best known of these is the VLSI Technology Research Association. This was established in 1976 as a four-year $200 million effort, with $88 million from government and $112 million from industry. It conducted advanced research in very large scale integration (VLSI) within its own facilities, the VLSI Cooperative Laboratory, an unusual arrangement for research associations. Its work is summarized in the Hudson Institute study mentioned earlier.[37]

By any standard, the VLSI program can be considered a success. It has produced over 600 patents and processes, and demonstrated the willingness and capacity of private corporations to work cooperatively under the aegis of a specially formed association, and with the technical and financial support of the government.

The latest Japanese effort in collective research is the Fifth Generation Computer Systems Project, known as ICOT. It is a concerted effort in both software and hardware to advance computer capabilities to a new order of magnitude that could represent certain elements of artificial intelligence (AI). ICOT was set up in 1981 with the intention of carrying out a 10-year program in which the government would invest $426 million. Eight corporations put up the seed money to start the program, and they provide some additional funding for current projects.

This program is unusual in several respects:

1. It has its own facilities, established in 1982 as the Institute for New Generation Computer Technology (ICOT).

2. It has a staff of 52 people, 42 of whom come from the sponsoring companies, government laboratories, and universities. Visiting researchers from other countries participate, and roughly two-thirds of these are software experts.

3. It is a consortium of eight companies and two national laboratories.

Participants in ICOT include the following companies and government laboratories. The asterisked names are also members of the Scientific Computer Research Association.

- *Industry*

    *Fujitsu Ltd.
    *Hitachi Ltd.
    Matsushita Electric Industrial Co.
    *Mitsubishi Electric Co.
    *NEC Corp.
    *Oki Electric Industry Co.
    Sharp Co.
    *Toshiba Corp.

- *Government*

> Electrotechnical Laboratory of the Agency for Industrial
>   Science and Technology (part of MITI)
> Nippon Telegraph and Telephone Public Corp. (NTT)

The program is an ambitious technical plan, laid out as broad major themes, such as basic software or advanced architecture, then in particular missions, such as knowledge base management system (within basic software) or relational algebra machine (within advanced architecture), and finally into a set of projects to achieve each mission. There are some signs that the objectives and the intensity of effort have been modified now that the program is into the third year. Nevertheless, it is a plan that, on paper, goes from R&D into a prototype system, with the output of all the component projects brought together under ICOT. To quote one observer: "ICOT is unusual in Japan in that it is a separate, independent, neutral organization that has been established to carry out a research project; the customary Japanese approach is to have each of the participating research institutions and companies conduct work on its own."[38]

The discussion in these few pages makes it clear that Japan does indeed make more effort than most other countries to concentrate technical resources on problems of national concern. There is more variety in the mechanisms used, and there is more attention to the involvement of private corporations in order to have the benefit of their inputs and improve the effectiveness of transfer and use.

Nevertheless, my conversations with friends and colleagues in Japan plus Americans with considerable experience there make it equally clear that this is a far cry from government direction of effort or control of technical activity. It is my understanding that the Japanese companies maintain thoroughly independent and competitive R&D programs. They may *choose* to join the various collective programs for particular technical activities, and to work in collaboration with the government in others. But those developments that are critical to their competitive status in Japan and on world markets are almost certainly confined to company laboratories. Japanese companies are not *that* different in their behavior than U.S. companies.

COOPERATIVE INDUSTRIAL RESEARCH

The growth of linkages and of collective research must not be equated to a decline of competitive behavior. There have been several important recent actions in collective industrial research activity in the United States. These new activities provide a mechanism for concentrating collective technical resources on selected areas in order to permit industrial corporations to focus internal resources on those efforts most critical for maintaining their own competitive status.

There is nothing particularly new about two or more companies cooperating in some technical activity. Trade associations are a well-established mechanism for such efforts, going back to the nineteenth century. These are more active in Europe with regard to technical programs than in the United States, since a substantial proportion of such European organizations have their own facilities, whereas in the United States the trade association operates most often by distributing funds for technical activities to be carried out by member companies, universities, even government laboratories.

By 1982 it became apparant to me and my colleagues at the NYU Center for Science and Technology Policy (CSTP) that there were some fundamental differences in the nature and functions of a number of newer collective industrial research organizations initiated in the preceding 10 years, and particularly since 1980. Further, we came to believe that these differences reflected some underlying pressures that were affecting the nature of industrial research. For these reasons, we initiated a study at the Center and, after some exploratory work, launched a more substantial study funded by the National Science Foundation.

Details of the data collected, the analyses, and discussion of the significance for the technical system appear in two references. One is the report to NSF of over 200 pages.[39] The other is an article in the *Harvard Business Review.*[40]

We obtained data on 59 associations representing 19 industry sectors. In 1984 these 59 groups supported about $1.6 billion of R&D, roughly 3 percent of all industry-funded R&D in the United States. However, the three largest groups accounted for $1.4 billion, leaving some $200 million distributed among the rest. While the study was intended to be representative, not exhaustive, no large groups were omitted, and probably few if any in the $5 to $10 million range, so the overall data very likely represents well over 90 percent in dollar value of R&D supported by U.S. collective industry associations.

These groups conduct a range of technical activities that fall broadly within R&D—research, development, testing, prototypes, and pilot plants—and some that are on the edge, such as technical information services, and related economic data. Many support graduate training, and some provide courses for a degree. Funds come almost entirely from corporate members, plus an occasional research grant from a government agency. A considerable amount of money is provided by the associations for technical activities at government laboratories. Three energy groups funded about $12 million at Energy Department laboratories in 1982, and a number of industry personnel are assigned each year at the National Bureau of Standards as part of their Industrial Research Associates Program at industry expense (there were 170 in 1982).

The collective research associations can be placed in many categories depending on their structure and functions. One classification that is useful is to define the following types:

- *Trade associations* which conduct both technical and nontechnical activities (for example, American Iron and Steel Institute)

- *Industry associations* formed specifically to conduct a research program either at universities (for example, the Semiconductor Research Corporation) or nonuniversity facilities (such as EPRI, the Electric Power Research Institute)

- *University-based centers* which were
  a. started with NSF funds and replaced by industry funds (such as the Polymer Processing Program at Massachusetts Institute of Technology) or
  b. funded from the start by interested companies (such as the Center for Integrated Systems at Stanford University)

- *Independent research institutes* to pursue nonproprietary technical advances (for example, the Sulphur Institute) or research related to the public welfare (for example, the Chemical Industry Institute of Toxicology)

- *Independent institutes* with dual focus on education and research (such as the Institute of Paper Chemistry, accredited to grant degrees, and the Textile Research Institute, affiliated with Princeton University)

- *Research corporations* which conduct R&D at their own facilities, funded by members, and which can lead to commercial exploitation for the benefit of the members (for example, the Microelectronics and Computer Technology Corporation and the American Welding Technology Applications Center)

What is there in this range of collective activity that suggests pressures acting upon industrial research? To answer this, I will place the various associations in three categories.

- Traditional trade associations founded in almost all cases prior to 1970
- Newer forms of organizations formed almost entirely after 1970
- Large collective organizations formed by regulated utility sectors

Regulated Utilities

The last category is related to the theme of this book with regard to the pressure to concentrate increasing technical resources. It is a narrow, though significant, category that is not related to the conventional competitive pressures on industrial research.

The regulated utility industries are technology-based, and in principle can derive great benefits from development of technical standards, improvements in processes and systems, and even major advances that can change underlying technologies radically. All of these actions can lower costs or provide better services. In the United States, utilities are private monopolies subject to regulations primarily of state and local governments. In Europe, these functions are normally conducted by government-owned agencies.

The principal mechanism for the process of technical change is industrial research conducted within the pressure of a competitive environment. Regulated utilities and government-owned monopolies do not possess that environment. How then can such organizations provide incentives for creative technical activity which, when successful, requires the introduction of change into a structure that is under no urgent competitive pressure to do so? The facetious answer is: with very great difficulty. One of the great accomplishments of the Bell Telephone Laboratories has been to maintain a research organization that was outstanding in both the quality and quantity of output within the regulated monopoly operations of AT&T before dives-

Table 9-3. Collective Research for Regulated Utilities, 1984.

Association	Technical Budget ($ millions)	Formed
Bell Communications Research	$878	1984
Electronic Power Research Institute	345	1972
Gas Research Institute (GRI)	114	1976

Source: C.S. Haklisch, H.I. Fusfeld and A.D. Levenson, "Trends in Collective Industrial Research," Center for Science and Technology Policy, Graduate School of Business Administration, New York University, August, 1984.

titure. This was accomplished largely by developing a system of peer pressure based on stature and recognition within the professional scientific and engineering communities. The recognition of those communities and the pride of technical accomplishment became critical driving forces, so that an internal pressure for technical advance permeated the organization.

For several major utility industries in the United States, a uniquely American solution to the problem of providing for technical change has been to create a separate R&D organization that is supported and controlled collectively by companies within that utility sector. The collective organization is separated from the technical operating needs of any company and the collective funding provides a base for independence in the selection and conduct of programs. Nevertheless, the board of directors and various technical committees are made up of representatives from the sponsoring companies. Thus, there are certainly inputs into the selection process regarding priority of technical needs from the companies, and there is at least the opportunity for technical coordination and technology transfer that can make the association's R&D most effective in terms of integrating technical advances within the industry.

The three most important collective research groups for the regulated utilities, and in fact the three largest collective industrial research activities in the world, are given in Table 9-3.

The newest and largest group, Bell Communications Research (BCR), which Bell also calls Bellcore, was set up as an outgrowth of the AT&T divestiture. It is owned by the seven regional phone companies created from the 22 operating companies of the Bell system. About 8,000 people were assembled within the BCR structure, half of them coming from Bell Labs. Much of the activity is devoted to networks and systems. A very major responsibility is the establishment of technical specifications for equipment, so that the regional companies, which are free to buy from any supplier, will all have equipment compatible with a nationwide system.

An important part of BCR is the Applied Research group with an initial staff of 500 people. Its principal research areas are in (1) mathematical and computer sciences and (2) solid state science and technology. It will pursue longer term work related to telecommunications networks and to new optical fiber technology, switching and signaling techniques, integrated circuits, and speech recognition. Based on the backgrounds of the people coming from Bell Labs and on the pressure from the operating companies to conduct competitive product developments separately, there is reason to believe that this research area of BCR will become progressively more long range and basic in character, despite the name "applied science."

BCR is of interest because of concern about the impact of divestiture on the Bell Laboratories. Bell Labs has been such a leader in the U.S. technical establishment that an adverse impact on it could mean an adverse impact on our technical base. It is far too early to judge, although Bell Labs must now be funded from a competitive enterprise with all of the opportunities and restrictions this implies. Still, any evaluation must take into account the new capabilities of Bellcore, which derives its funds as part of the rate base charged to phone users.

The two "older" collective R&D activities of regulated utility sectors are the Electric Power Research Institute (1972) and the Gas Research Institute (1976). They act to advance the process of technical change for their respective sectors. EPRI has 489 members, GRI 210. Both support a range of technical activities, concentrating on applied research and development and extending into prototypes and pilot plant operation. EPRI contracts work to industry, to universities, and to government laboratories. GRI also supports some work at government laboratories and at universities. Much of its research activity is conducted at the Institute for Gas Technology (IGT), located at the Illinois Institute of Technology in Chicago.

These three collective organizations of regulated utilities are trying to advance complex technologies that require substantial resources for R&D and its applications. They are able to provide these resources by having the R&D expenditures absorbed in the telephone, electric, and gas bills of the consumer. They are, in a sense, quasi-government activities pursuing technical advances for public objectives, but capable of planning and implementing programs with some independence and objectivity. It is an American approach.

There are some lessons for public policy in having large research efforts funded within a broad umbrella. It is possible to consider some analogous mechanism for basic research. However, such possibilities require very careful examination, since the lack of line-item funding often loosens accountability and feedback. In the three cases discussed here, those elements are provided by representatives of some very profit-minded organizations, regulated or not.

The regulated utilities form a special group. The collective research organizations they established tell us something about the concentration of resources to achieve technical progress in complex industries. They tell us little about recent pressures in industrial research. For that, we must examine the differences between the older trade associations and the newer collective organizations.

Collective Industrial Research in Competitive Industries

The earlier groups, certainly those formed before 1970, were concerned with noncompetitive activities. These included basic research, safety, health, data collection and dissemination, testing, development of measurement methods, and so on. Training and support for universities to assure supply of competent graduates were present in many of the groups.

Any of these activities could have been conducted by the medium size or larger companies that belonged to these associations. The technical requirements were within the resources available to these companies. The reason for doing them through a collective mechanism was primarily as a cost-effective measure. There was occasionally the factor of credibility in a group effort that would not be taken as self-serving, for example, when data on particular materials or systems is published.

Many members of these associations were smaller companies which did not conduct technical activity other than that necessary to produce and test their own products. In other cases, an entire industry might have very little technical effort. The industry might be composed largely of small companies. The work of the collective group in these instances provided technical efforts as a substitute for individual activity, rather than as a cost-saving supplement.

In general, these original associations had, and have, fairly modest budgets. In 1982 only three had a budget of over $10 million, while 29 had a budget of under $1 million. Much of the technical programs were conducted by those with budgets between $1 million and $10 million. The collective research effort provided a technical infrastructure and pursued modest improvements in products and processes.

The newer collective industrial research organizations possess quite different characteristics. Their budgets are likely to be over $10 million, particularly in electronics. In 1984 MCC had a projected budget of $50 million and the Center for Integrated Systems at Stanford estimated $30 million in expenditures. SRC was at $12 million and growing, while the Microelectronics Center of North Carolina (MCNC) expected to spend $7 million.

An important distinction lies in the technical area and industry sector. The older groups were heavily in commodities and mature industries. They were in paper, metals, textiles, foods. The technologies were changing slowly, and the collective groups acted for the industry to strengthen R&D, to attract graduates, and to develop new markets. Individual companies in these fields were not technology intensive, with R&D expenditures roughly 1 percent of sales or less.

The newer groups represent wholly different disciplines and corporations. They are in rapidly changing technical fields—computers, semiconductors and, to a lesser degree, biotechnology. The chemical industry is involved actively. (The chemical industry was also active in earlier trade associations, including the oldest one, formed in 1872, the Chemical Manufacturers Association.) These are all highly technology intensive industries, with R&D expenditures roughly 3 percent of sales for chemicals, 7 percent for computers, and 8 percent for semiconductors.[41] Moreover, the corporations participating in the newer groups are among the largest corporations with major R&D organizations that are in the forefront of U.S. technical advance.

Not surprisingly, then, the functions of the newer groups are different. They tend to emphasize those early stage technical activities from basic research to generic forms of technology, all of which are being described as "precompetitive," to use the phrase put forward for ESPRIT and the Alvey Program in Europe. Occasionally, the phrase "procompetitive" is used, which implies more action orientation. There are accompanying functions of graduate training, but there too the objectives are different. The older trade associations wanted to induce competent people to seek careers in their industries. The newer groups are faced with current or potential personnel shortages in specific areas.

The new groups are conducting or supporting R&D that is needed to provide significant technical advances in the industries affected. These activities contribute to developing the technical foundation for the next generation of products and processes. Ten or twenty years ago, much of this activity would have been carried out within the company laboratories. Each would have taken some different approach, some different part of the technical frontier, developed its own foundation, then gone on to whatever products or processes would be based on that foundation and would fit its own business plan.

Why have the new groups been entrusted with some of these activities today by companies such as IBM, RCA, Honeywell, Intel, and the like in the United States, or Philips, Siemens, Thomson, and ICL in Europe? Simply put, the resources required to advance the frontiers individually, then conduct the very complex and large-scale developments necessary to create a competitive niche, may not be available to even the largest corporations in an industry subject to rapid technological change. Sophisticated, high-tech industries in which the leading corporations are technology intensive require ever-increasing resources to maintain the competitive status of each corporation. As Carlo De Benedetti, chairman of Olivetti, said with regard to Olivetti's joint venture with AT&T: "We can't develop everything ourselves."[42] And John Young, president of Hewlett-Packard, commenting on the Center for Integrated Systems at Stanford, summed up his firm's position: "Even a company as large as Hewlett-Packard, with half a billion dollars to spend on R&D every year—and IBM, with much more—sense that this kind of cooperative effort is going to be necessary for us to succeed in the competitive international environment between now and the end of the century."[43]

Does this mean that any of these corporations have delegated their competitive R&D to the collective research groups? Not at all. They use the words *precompetitive* and *procompetitive* somewhat loosely, but the effect seems clear. If one considers all the technical efforts called for by the continuing process of technical change, the earlier and more generic phases are less critical to the competitive status of the corporation than the later, more product-oriented phases. Some corporations may delegate all of the most basic and generic research, while an IBM or Philips may delegate some of these activities. All corporations maintain those activities necessary for their competitive status as defined by their individual strategic business plan, which is by definition compatible with their internal technical resources.

It is of some interest to note that differences in collective industrial research within the biotechnology industry contrasted with electronics. There are only a few formal organizations which conduct R&D collectively in a public structure similar, say, to the Semiconductor Research Corporation. Some programs in biotechnology and related areas are undertaken by several of the established associations, such as the Pharmaceutical Manufacturers Association Research Foundation. There is the Center for Biotechnology at Washington University, set up in 1983, supported by a few corporations, which may lead to a number of similar groups in the future. But there is no collective research organization set up and operated by member corporations similar to SRC, or established at universities on the scale of the Microelectronics Center of North Carolina, or the Center for Integrated Systems at Stanford.

Microelectronics and biotechnology are both rapidly changing high-technology areas, involving corporations which are large, technology intensive, and conducting very substantial R&D efforts. Yet one pursues technical advances through collective organizations, the other does not. Why the different behavior?

The answer is twofold. First, the disciplines are at different stages of maturity with respect to converting technical advances into commercial use. Second, and surely related to the first, there is indeed collective industrial research going on in biotechnology, but it is taking place through financial mechanisms rather than associations.

Biotechnology is at a young and highly competitive stage. Patentable processes and know-how are of great importance commercially. Opportunities for cooperative R&D where the results can be shared are difficult to identify. This period is very like the growth period in semiconductors 25 or 30 years age.

Numerous privately structured activities in biotechnology have been organized which combine R&D activity with a business plan for exploiting the results. Concentration of technical resources occurs through financial investments that can involve one or more companies with private investors. One example is Engenics, formed by six corporations to support both basic research and developments in biotechnology. It had an initial investment of $20 million for four years, and established its own facilities to conduct development work.[44] The outputs would be commercialized, possibly by one or more of the investing companies. Basic research was conducted though its Center for Biotechnology Research. This provided funds for research projects to be conducted primarily at Stanford University and the University of California at Berkeley. The Center had a budget of $730,000 in 1984. Engenics is actually a research corporation, similar to MCC but at a lower funding level.

A more frequent activity in biotechnology has been the use of R&D limited partnerships (RDLPs), which will be discussed in Chapter 13. Investors, including corporations, can share the costs of an identifiable R&D program or group of programs. There are tax incentives for pooling financial resources, and the partners then have the opportunity and challenge of developing income from the patents and know-how that result. Genentech has used this mechanism successfully to follow different directions suggested by their research in genetics. Each limited partnership has included a major corporation which can carry through applications of successful research in their own field. The actual R&D is conducted by Genentech, paid for by the funds of the partnership.

This financial mechanism has been used to leverage specific internal research programs, such as by Becton-Dickenson.[45] Many smaller applications have been financed by investment houses or private investors. While the use of RDLPs has covered a wide range of technical areas, it has been particularly heavy in biotechnology. The combination of major scientific advance, widespread potential for commercial applications in major industries, yet still very much in the development stage appear to present a particularly fruitful field for the RDLP. It has attracted considerable funding from many sources to pursue technical advances.

COOPERATION VERSUS COMPETITION

Recent years have seen the creation of technical linkages throughout the world—joint ventures among two or more companies, national and international programs of technical cooperation, new forms of collective industry research association, even new financial mechanisms to provide collective forms of funding for R&D. The considerable mass of data and examples given in this chapter probably include most of the best-known and largest of these efforts. Numerically, they are only a fraction of the substantial and increasing activity in such linkages.

I have stressed here primarily linkages which involve private corporations. We expect research cooperation among universities, among similar government agencies in different countries, and under the umbrella of professional societies. The examples given have referred to several of these traditional sectors, and have pointed up the initiatives taken in collective programs that are conducted primarily by corporations to bring in the efforts of universities and government agencies.

Yet the thrust of collective research activity in the past few years has been among private, competitive corporations. It represents a joining of resources—technical and financial—to pursue technical advances in areas of common interest. However long-range, basic, or generic these programs may be, they are no different in their nature from programs that would have been conducted by the companies individually 10 or 20 years ago, or that are being conducted by many of the participating companies today.

Do these collective industrial research activities represent some form of research cartel that acts as a common source for all the companies in their product developments? Have the companies delegated their competitive growth to collective R&D so that similar directions for new developments will result? The answer is, most emphatically, no. There is not the slightest indication that the individual corporations look to the collective research activity for initiative, for decisions, or for critical technical leadership that will determine what the company does in the future. There is every indication that the collective research programs are in addition to, not instead of, corporate R&D. There has been no resulting cutback in industrial research expenditures.

How do we reconcile the growth of cooperation in R&D with the continuing vigor of competitive efforts by the participating companies? The simplest explanation is that these many forms of technical linkages offer an increase in resources available to the industrial corporation. Each company, in pursuit of its overall corporate strategy, will decide what mix of technical resources is available to implement that strategy—internal, external, or collective. Indeed, given the growth of collective activity and the increased choice of resources this creates, new options for corporate strategy arise.

An intriguing aspect of all this is the international nature of many of these linkages, discussed in the next chapter. They can lead to new forms of competitive behavior. My colleague Carmela Haklisch, who has been conducting research into much of these recent developments, has a favorite saying: "The new patterns of cooperation will determine the new patterns of competition."

An increase in common technical efforts and in corporate partners to exploit them can create the base for more vigorous competition. This is certainly true if, without the cooperation, technical progress would be slowed. The growing sense that even increased R&D budgets within the largest companies are not adequate to create the technical progress needed for growth is certainly an important force in support for collective research activity. The perception that other companies, other countries, other regions were strengthening their technical base through such activities is an additional force.

There is indeed a constant pressure to increase technical resources continually in order to maintain our rate of technical change. When these increases create too much pressure on individual corporations, other mechanisms come into being. That is indeed what we are seeing today.

Collective activity is not lessening competition. It is strengthening the worldwide technical base. This is how we should interpret the words *precompetitive* or *procompetitive* used to describe ESPRIT, Alvey, and the MCC. The spectrum of technical activities embraced by industrial research ranges from noncompetitive public activities, to basic research, to engineering and prototypes. The boundary for collective activity has simply moved a bit closer to the practical end of the spectrum.

NOTES TO CHAPTER 9

1. *Business Week*, May 27, 1985.
2. *Wall Street Journal*, May 22, 1985.
3. *Wall Street Journal*, June 5, 1985.
4. *Wall Street Journal*, June 6, 1985.
5. "Machines That See Look for a Market," *Fortune*, September 17, 1984.

6. *Fortune*, July 23, 1984, p. 8.

7. "Will Ford Beat GM in the Robot Race?" *Business Week*, May 27, 1985.

8. *Wall Street Journal*, April 30, 1985.

9. *Business Week*, May 27, 1985.

10. *Wall Street Journal*, May 10, 1985.

11. "From Autobahn to Aerospace," *Business Week*, May 20, 1985.

12. "IBM's Italian Connection in Factory Automation," *Business Week*, October 8, 1985.

13. "Corning Glass Shapes Up," *Fortune*, December 13, 1982.

14. "For Chipmakers, National Boundaries Begin to Blur," *Business Week*, May 6, 1985.

15. *Wall Street Journal*, Mqrch 7, 1985.

16. "Two-Way Street," *Fortune*, January 23, 1984.

17. *New York Times*, May 15, 1985.

18. "Japan's Two-Fisted Telephone Maker," *Fortune*, June 25, 1984.

19. David Dickson, "Europe Seeks Joint Computer Research Effort," *Science* 223 (January 6, 1984): 38–30.

20. Nico Hazewindus, Director, Corporate Product Development Coordination, N.V. Philips, Conference on Fifth Generation Computers, London, May 25, 1983.

21. "Importing Science," *Wall Street Journal*, October 5, 1983.

22. Lois Peters and H.I. Fusfeld, "Technical Exchanges between U.S. and Japanese Industry," Center for Science and Technology Policy, Graduate School of Business Administration, New York University, 1986.

23. Mary Damask, "Industrial Association Utilization of Federal Laboratories," in the report "Trends in Collective Industry Research," edited by C.S. Haklisch, H.I. Fusfeld and A.D. Levenson, New York University, August 1984.

24. Lionel Boulet, "Overview of Energy Policy," in *Industrial Productivity and International Technical Cooperation*, edited by C.D. Haklisch, and H.I. Fusfeld (New York: Pergamon Press, 1982).

25. Commission of the European Communities, 17th General Report on the Activities of the European Communities, 1983, p. 51.

26. For a detailed review of ESPRIT, see C.S. Haklisch, H.I. Fusfeld, and A.D. Levenson, "Trends in Collective Industrial Research," Center for Science and Technology Policy, Graduate School of Business Administration, New York University, August 1984.

27. *Bulletin of the European Communities*, Supplement, May 1983, p. 29.

28. *Wall Street Journal*, March 14, 1985.

29. June 4, 1985.

30. "West Germany Plows Major Technology Investment," *Science* (March 30, 1984): 1375.

31. *Sweden Business Report*, November 11, 1983.

32. *Alvey News 1* (September 1983), joint publication of the Institution of Electrical Engineers and the British Computer Society.

33. David Dickson, "Britain Rises to Japan's Computer Challenge," *Science* 220 (May 20, 1983): 799.

34. Detailed discussion is given in C.S. Haklisch, H.I. Fusfeld, and A. Levenson, "Trends in Collective Industrial Research," Center for Science and Technology Policy, Graduate School of Business Administration, New York University, August 1984, pp. 184–195.

35. Roy Rothwell, "Trends in Collective Industrial Research," TNO, Delft (Netherlands), 1979.

36. J. Wheeler, M. Janow, and T. Pepper, "Japanese Industrial Development Policies in the 1980s. Implications for U.S. Trade and Investment," Hudson Institute, Research Report HI–3470–RR, October 1982, p. 155.

37. Ibid.

38. "ICOT: Japan Mobilizes for the New Generation," *IEEE Spectrum*, November 1983.

39. C.S. Haklisch, H.I. Fusfeld, and A.D. Levenson, "Trends in Collective Industrial Research," Center for Science and Technology Policy, Graduate School of Business Administration, New York University, August 1984.

40. H.I. Fusfeld and C.S. Haklisch, "Cooperative R&D for Competitors," *Harvard Business Review 63*, no. 6 (November–December 1985): 60–76.

41. "R&D Scoreboard," *Business Week*, June 20, 1983, pp. 122–153.

42. *Wall Street Journal*, October 5, 1983. Nico Hazewindus, of the Philips Company, has made remarks along the same lines with regard to Philips' participation in the ESPRIT program.

43. Frederick H. Gardner, "The Center for Integrated Systems," *Hewlett-Packard Journal*, November 1983, p. 24.

44. Unpublished report to the National Institutes of Health.

45. *Business Week*, "Laying off R&D Risks on Tax Shelter Investors," March 5, 1984.

10 THE ECONOMIC CONSEQUENCES

Each organization that generates science and technology is becoming increasingly affected by, and possibly dependent on, external sources of technical activity. This results in linkages of all sorts.

The fundamental reason for the linkages is that technical change is complex and costly and the magnitude of the technical enterprise today is great. A corporation whose strategic growth plans require it to generate technical change finds its internal resources strained or inadequate. Even one whose growth can be achieved by adapting and modifying technical advances, by creative design and engineering, and by generating modest technical advance must draw upon an increasing range of science and technology and is faced with a greater number of competitors. In either case external technical linkages are an economic necessity.

The pressure to communicate technical information, to transfer technology, to combine technical resources are no respecters of national boundaries. When those boundaries place restrictions on these linkages or on the technical flow, these restrictions limit the effectiveness of the process of technical change.

The networks that would be created within the worldwide technical enterprise among public and private organizations, governed only by a competitive market system uninhibited by national boundaries, would represent an optimum system for the process of technical change. Conditions are imposed on this ideal system due to individual nations' legitimate concerns with their own security, employment, health and safety, natural resources, industrial policy, and national strategy and culture.

Almost by definition, national boundaries decrease the maximum potential of the technical enterprise for economic growth. Again by definition, each country evolves within its boundaries a technical system—private and public—which represents a balance of political decisions with regard to *all* the national objectives that depend upon it. The important question is whether each country recognizes the compromise which it imposes upon the technical system by calling upon it to provide for economic growth in combination with the other national objectives.

Inevitably a tension is created between private and public sectors. The process of technical change creates pressures for the internationalization of R&D—not only research but development. The principal burden for pursuing this internationalization falls on the private sector as a result of the competitive pressure for generating and using technical change. The private sector has the responsibility for converting technical advances to economic use. Hence, any national objectives that limit the internationalization of R&D limit the economic potential of its domestic industries.

MAINTAINING INTERNATIONAL COMPETITIVENESS

It is critical to clarify the exact meaning of "international competitiveness" because (1) our technical and commercial activities are all international in nature and (2) important elements of national policy are being developed on the basis of international competitiveness. The new NSF program for Engineering Research Centers, proposals to modify antitrust laws, the formation of collective research activities such as MCC and SRC, the stimulation of R&D through financial incentives—these and more are advocated in great part because of the need to improve U.S. international competitiveness. It is the basis for such national and regional programs as Alvey and ESPRIT.

Once upon a time, the world was simple and the meaning of the phrase *international competitiveness* was clear. A company was owned by citizens of the country in which it was headquartered and its products were manufactured in the same country. To the extent possible, those products were then sold abroad. If the company did so successfully, it was internationally competitive.

Of course, that description is incomplete. The materials used to manufacture the products may have been brought in from other countries. They may have been imported in various forms from truly raw materials to semifinished goods. Financing may have come from sources in other countries. The technology used in manufacturing resided in the minds of the employees, who occasionally were brought in as immigrants for that purpose. That was technology transfer!

Today's world is far more complex, and we have to consider separately each factor contributing to the system by which products are designed, manufactured, and sold in international markets. There are at least three separate questions about the process:

1. Who owns the enterprise?
2. Where are the products manufactured?
3. Who is generating the technical base?

These are the principal elements of international competitiveness: ownership, production, and technology. A leading position in one element does not necessarily dictate the choice of the other. The location of manufacturing and assembly facilities is determined normally by economic considerations. The generation of new science and technology takes place where there is an adequate concentration of technical resources plus an appropriate R&D environment. Ownership is presumably the element that is based on financial resources and is most flexible with regard to geography, although there may be more barriers to the flow of money than the flow of technology.

Suppose the MCC is highly successful in generating major technical advances in microelectronics and computers. The U.S. companies which participate acquire the leading position in particular products and systems based upon those advances. Suppose further that those companies, in order to develop the widest possible markets and compete with each other, establish manufacturing facilities for those products and systems in Korea, Hong Kong, and Taiwan. Have we become internationally competitive in microelectronics?

Consider the opposite example. Suppose the ESPRIT program is highly successful along with the national programs of England, France, and West Germany in the fields of microelectronics and computers. The European companies now possess the leading position in a range of products and systems based upon the superior science and technology emerging from their programs. In order to exploit the U.S. markets, and to compete with each other there, the major European electronics companies expand production facilities in the United States. Is the United States now internationally competitive in microelectronics?

The examples only point up the need to think clearly. They are unrealistic, not because they are improbable, but because they ignore both the complexities and the fundamentals in the process of technical change. Some of these are explained in the following paragraphs.

1. Any advanced product or system integrates many components into its assembly. The value added by the producer of the final product can be only a fraction of the total value. A large electronic system, a jet aircraft, an automobile will bring together materials, components, and subassemblies that are developed and produced in other countries.[1] Two-thirds of each personal computer assembled and sold by IBM is made in Japan or Singapore, while half the components of American machine tools are manufactured overseas.[2] The location of the assembly operation is only one source of employment in the total production.

2. The establishment of an effective and competitive manufacturing process that implements the development of complex high-technology products must be done in close cooperation with the technical organization. The initial manufacturing is therefore very likely to be performed at a location convenient to the R&D or at least with arrangements to have some of the R&D personnel assigned to the process of placing the manufacturing facility in operation. Only after the primary manufacturing plant is successfully under way, usually in the same country as the R&D organization, would additional or supporting manufacturing be initiated in other countries.

3. The more advanced, the more complex, the more costly a major new product or process, the greater the probability that it involves technical linkages among two or more companies—licenses, joint venture, or technical cooperation in some other form.

4. Most important, the conduct and growth of R&D requires adequate markets, revenues, and profits. Placing manufacturing plants in another country as a competitive necessity may be a major economic contribution to strengthening R&D in the home country.

In brief, the concept that a single company can develop and control all the science and technology embedded in an advanced product or system, and can then clearly separate the technical base in one country from a complex manufacturing location is another, is rarely valid or practical. The large number of linkages already in existence and the continuing growth of such arrangements create patterns and networks that quickly contradict any simplified view of international industry practices or of international competitiveness.

What we want depends on who answers the question. When a political leader or labor leader expresses concern, the priority is jobs. Within reason, labor's distress in the event of Japanese dominance of the technical advances underlying microelectronics would be minimal if extensive manufacturing and assembly were conducted in the United States. The opposite situation would surely be more distressing to them, that is, if U.S. companies maintain technical leadership by carrying out the bulk of their manufacturing and assembly outside the United States.

The concern of government officials responsible for national security is first with the nation's capacity to produce technically superior military defenses and weapons. The principal objective is to achieve and maintain domestic leadership in critical fields of science and technology plus the capacity to convert that leadership into usable devices and systems. Total employment for commercial exploitation is secondary as long as there is adequate manufacturing to ensure the required production and the availability of skills.

Those officials and economists who focus on trade and balance of payments, for example, in the Departments of Treasury and Commerce, are concerned with actual cash flow. Payments for licenses and know-how, capital investments, payment of profits are all relevant factors, in addition to specific imports and exports. The decisionmakers of multinational corporations are critical in these items.

Given these several sets of concerns and the complex interactions that characterize the process of technical change today, what is a reasonable position to achieve with regard to international competitiveness? It would seem that any such position must take account of the interdependence of R&D and of industry generally across national boundaries. Any approach to international competitiveness that is based on the ability to isolate, to exclude, to be completely self-sufficient—however desirable that may be in critical industries and in time of great stress—must realize that this runs counter to the effective operation of our technical enterprise and of international business. The cost and inefficiencies that such actions would impose on U.S. industry, if indeed such an approach is possible at all, would very likely overshadow any perceived advantage.

Three objectives would take account of legitimate national concerns in order to achieve a desirable level of international competitiveness:

1. Reasonable balance within the United States between domestic and foreign decision-making in industries important to GNP, and between U.S. corporate decision-making abroad and foreign corporate decision-making in the United States

2. Economic competitiveness of domestic manufacturing in a reasonable number of industries for both domestic and world markets

3. Maintaining the capacity to initiate and produce change in present products, processes and services, and provide a base for future business development

The third objective is the key to where a country is positioned with regard to international competitiveness. At least it is the major new element among the traditional factors in world trade. Thus, as an indicator of what leverage a country may possess in this area, we can ask several questions about the capacity for technical change:

-   Is there technical competence in the forefront of some or all of the principal industrial sectors which are important to the country's economic base?

-   Is there the technical capacity to produce competitive advances in these industrial sectors?

-   Is there the technical capacity to generate the foundation for major change in currently important industries, or to establish the foundation for new industries?

-   Is the total technical enterprise of the country effective in the conversion of technical change to economic use?

Positive answers to these questions mean that a country has at least some options in international competitiveness. We may *choose* to take licenses for use of outside technology, or to import components of manufactured systems; we may *decide* to conduct manufacturing or assembly in other countries; we may *allow* foreign corporations to conduct manufacturing or assembly in this country. There is some freedom of action in economic affairs when choices are available from a solid base of competence in generating science and technology.

This brings us back to the growth of technical linkages presented in the preceding chapter. The many actions that have been taken arise in general from the needs of individual corporations, facilitated in particular cases by national and regional programs. Each corporation makes use of these external linkages to strengthen its technical capabilities and thus its individual competitiveness in the world market. The aggregate improvement in the position of all corporations within a single country clearly improves the international competitiveness of that country from whatever viewpoint. The greater technical base and the additional options resulting from the use of technical linkages results in the country being in a better position than it would be if each of its corporations restricted their growth only to their internal technical capabilities.

Thus, once technical linkages become an important mechanism for expanding technical resources, as they have, there is great pressure for all technically based companies to pursue these where relevant. The senior technical officer of one of the largest companies in Europe, with major research laboratories, told me in 1984: "My company objects in principle to the ESPRIT program, because we feel the electronic companies may lose some of the sharpness in the competitive thrust of their R&D. Now the EC wants to establish a comparable program in biotechnology that affects our own research, and we are equally opposed. But if they actually set up such a program, of course we will join it. We would just have to maintain our own competitive position!"[3]

Thus international competitiveness is affected by the linkages within the technical enterprise and the participants appear to be strengthened. These linkages result in effects that influence international competition. They provide mechanisms for technology flow that tend to create a roughly common level of technology throughout the industrialized countries of the OECD. The specific linkages create networks which can provide the basis for business partnerships and joint ventures which follow through in marketing and production those technical advances resulting from the linkages based on internationalization of R&D.

As we examine these networks formed within the technical enterprise, we become very much aware of the fact that competition occurs directly between companies and only indirectly between countries. Those technical linkages which are effective are those which strengthen the individual corporate participants, which can then build upon the technical advances to develop products that fit their strategic growth plans. As companies within a country grow stronger, the country's position in international trade strengthens.

As the international character of the networks makes obvious, the competitive patterns that emerge begin to blur national boundaries. Combinations of companies from different national bases act to pursue jointly the business opportunities generated by their technical cooperation. The cause and effect may be reversed, since the strategic plans of a corporation may include the active cooperation of foreign companies first to provide some important technical element of the plan, and second to conduct some portion of the business activity resulting.

One recent case illustrates how strategic plans create international linkages. The Nippon Electric Company (NEC) has developed its own line of microprocessors (micros) to break away from dependence on licenses from U.S. companies. Its plans for overall systems development and subsequent marketing are described by *Fortune* magazine as follows:

> the six micros NEC has announced in the V series blankets Intel's line of micros and for the most part are compatible with them. NEC is going for broke. It has lined up Zilog in the U.S. and Sony in Japan as back-up suppliers, introduced 11 peripheral chips to go with the V series, teamed up with Hewlett-Packard's Japanese subsidiary to make a development system, and signed deals with a gaggle of U.S. software houses.[4]

The more novel and far-reaching examples are of those patterns that arise from the technical linkages. Consider the networks that have arisen in microelectronics and computers.[5] Figure 10-1 shows the companies within the ESPRIT program, the Alvey Program, and several of the subcombinations that some of the companies have created. Figure 10-2 provides some of the patterns that emerge from the actions of U.S. companies through their participation in SRC, MCC, and the very high speed integrated circuits (VHSIC) program,

Figure 10-1.   Principal European Linkages.

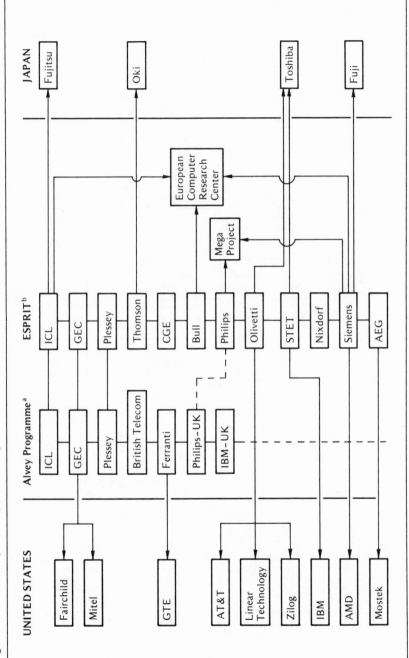

a. Participants in one or more major programs.
b. Original participants.

Figure 10-2. Principal U.S. Linkages.

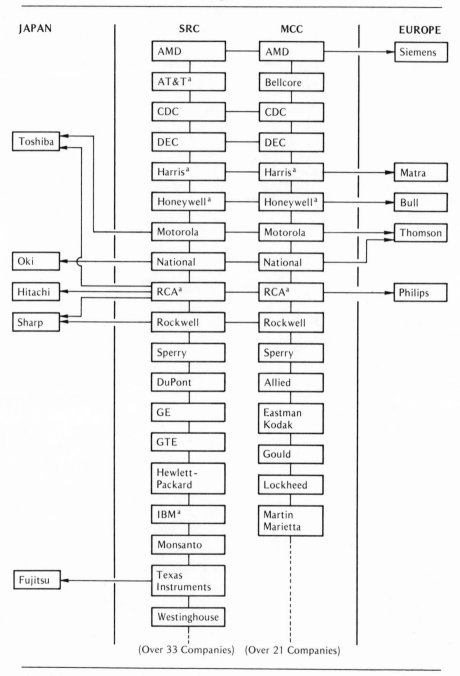

a. Prime contractor in VHSIC.

with a number of the subcombinations in which those companies have joined. An important linkage shown is the agreement with specific Japanese companies to be second-source suppliers for a number of U.S. semiconductor producers.

The one strong message of these diagrams is that the process for generating new science and technology, providing mechanisms for the flow of results, and converting them to economic use is complex and is increasingly international in character. Any national policies regarding international trade or the flow of technology that do not account for the functions provided by the linkages will not only fail in their objectives but may do positive harm to the technical enterprise within that country.

ECONOMIC THEORIES OF TECHNICAL CHANGE

Economists have been concerned with technological change at least since the work of Adam Smith. For this century, Joseph Schumpeter stands out as the main proponent of the view that technological change is the driving force of economic growth.[6]

Economics is concerned largely with technical change at the macroeconomic level. A seminal contribution in macroeconomics was Robert Solow's 1957 attempt to locate the sources of economy-wide productivity growth in the period 1909 to 1949.[7] Using the concept of an "aggregate production function," Solow discussed that he could explain only 10 percent of this growth in per capita output by increases in the amount of capital employed per worker; the residual 90 percent, he reasoned, must reflect a "shift" in the production function itself—a shift he associated with technological change. Edward Denison undertook a similar study that refined Solow's method.[8] Yet, after including more factors that might have accounted for increased productivity, Denison found that "the advance of knowledge" still accounts for 40 percent of the growth of output per worker in the period 1929–1957.

Economists have also studied technological change at the microeconomic level. The work of these economists is much more closely related to the issues of organization and processes that I have been addressing.

This corps of scholars is growing, and their work is proving highly valuable to the practitioner and the noneconomist. Among the principal members of this group today are Edwin Mansfield at the University of Pennsylvania; Richard Nelson at Yale University; Nathan Rosenberg at Stanford University; Christopher Freeman and Keith Pavitt at Sussex University (England).

Mansfield's approach is essentially to apply at the firm or industry level the techniques that Solow and Denison used at the economy-wide level. In one study Mansfield looked at interindustry and inter-firm differences in the rate of productivity growth. He found that those industries and firms with the higher R&D intensity tend to have the highest rates of increase in productivity.[9]

Another emphasis of Mansfield is his work on the "social return" from R&D. This deserves some detailed consideration because it has been used as a basic concept in justifying public policy in support of government financing of R&D activity.

The reasoning is straightforward. The decision to invest in R&D for economic purposes is made by the individual firm. Each company has some minimum rate of return which it should expect to receive before it proceeds with the R&D investment. This would be typically in the range 15 to 20 percent for a program of medium risk, a bit less if the program is a modification of some existing product or process, and a lot more if the program represents a major technical advance with considerable associated risk.

Mansfield's analysis, based on case studies of individual innovations, shows that the rate of return of a successful innovation to society is typically greater than its rate of return to the company introducing the innovation. A new process can provide additional market share by lowering the cost of a metal or agricultural product by a few cents per pound, but society benefits from the total reduction in price of all pounds sold by all companies. A new product that generates profits for the company through added manufacturing value stimulates related products by other companies. Whatever the technical change that produced a specific return for the company making the initial investment in R&D, the cumulative return to society from lower prices, added production, and so on, is much greater. Mansfield's study of 17 innovations concluded that the median rate of return to the companies was about 25 percent before taxes, while the median return to society was about 56 percent.[10]

The precise numbers are not critical, but the concept leads to some interesting conclusions. There is always a cut-off point in the expected rate of return below which the individual company will not proceed with the R&D. Suppose a company has the option of carrying out a program which has a probable return of only 11 percent, while the return to society would be about 25 percent. The company would very likely decide not to proceed, thereby depriving society of a 25 percent return.

The broad conclusion from this analysis is that competitive firms tend to underinvest in R&D with regard to total return to society. This is the best theoretical justification for policies to stimulate increases in R&D. However, it does not follow that any mechanism that would increase R&D effort can be justified, although this reasoning is occasionally present in arguments for legislation or appropriations intended to raise the national level of R&D expenditures for the civilian sector.

Mansfield demonstrated that the value received by the corporation from an investment in R&D is less than the value received by society as a whole. That does not mean that "society" can increase R&D investment outside the private sector and expect to receive anything like a proportional return. As repeatedly stressed throughout this book, (1) the many benefits received by society from technical advances are enhanced and expedited through an organized system for producing technical change; (2) our modern system of industrial research has provided a balance between R&D and the capabilities for exploiting results; and (3) the selection of technical products, the feedback from manufacturing and markets, and the investment of capital all work coherently to produce economic value.

The public awareness of benefits from technical change and the desire to increase these benefits have on occasion led to an unconscious reversal of Mansfield's conclusions. Recent discussions about the emphasis on defense research, for example, have introduced comments that the government ought to spend more on civilian research instead. Implicit in such statements is the belief that any increase in R&D effort, whether at random or by decision of government, will produce economic gain. While there will surely be some benefit, the analyses of Mansfield, and of Schumpeter before him, referred to the results of purposeful organized R&D integrated within a balanced industrial system. We cannot disconnect the R&D from that system and expect the same relationships with economic growth to hold.

Another valuable and thoughtful commentator is Nathan Rosenberg of Stanford University, an economic historian and contributor to the economics of technical change. His most recent work has included the growing internationalization of technical advances in industry. One aspect of Rosenberg's thought of special relevance to the themes of this book is his concern with innovation in the largest sense. While many researchers, writers, and the general public focus on major innovations, Rosenberg points out that real-life technical change occurs through many small steps.[11] "Incremental innovation" occurs steadily, and innovation is often most significant in its economic impact when it represents some modest process improvement. Furthermore, understanding the cumulative economic value of an innovation, even an unglamorous step forward, consists of tracing the diffusion of the specific technical advance through various industries and accounting for the economic benefits received by each of those industries.[12]

Thus, Rosenberg's work shows that economic value is derived at all levels of technical change. Taken together with Mansfield's studies, his work provides a general analytical base for viewing R&D as an investment decision. This sets out some criteria for allocating resources to R&D. The allocation is carried out by the individual corporation, and such private decisions are themselves one benchmark of the proper level of R&D in society. There may be other social objectives that require a different allocation of R&D. In those cases, government decisions are more appropriate. We should always be clear as to which objectives are being considered when R&D levels are discussed.

The last topic—the appropriate role of government in R&D allocation and in setting a middle course between undirected basic research at universities and integrated R&D within the industrial system—has been considered extensively by Richard R. Nelson of Yale University. His early contributions included a classic volume that studied government role in a handful of large-scale programs and in a limited number of specific industries.[13] Later, he analyzed the poor record of government on "picking winners" with regard to R&D priorities, thus setting forth criteria for a coherent industrial policy.[14]

One additional theory of interest holds that technology investment and economic growth (or decline) are all related through a long-term business cycle. This was an idea that Schumpeter advanced in his massive 1939 study of the business cycle.[15] He identified and named a number of "waves," the most famous of which is the "Kondratieff wave," after the Russian economist N.D. Kondratieff.

The latter's thesis that a 50-year cycle or "long wave" characterized capitalist economics ran counter to the Marxist doctrine of an eventual capitalist decline rather than resurgence, and earned him exile to Siberia in 1930.[16]

The "long wave" theory was spelled out by Forrester at Massachusetts Institute of Technology more recently in terms of four phases: Phase 1 is a 15-year recession period; phase 2 a 20-year reinvestment and growth period; phase 3 is a 10-year overbuilding period; and phase 4 a 5- to 10-year period of turbulence, ending in recession. Roy Rothwell of Sussex University has argued that there have been several such waves in the industrial era.[17] He specifies four such innovation-driven cycles as follows:

1st wave	1782–1845	Steam power and textiles
2nd wave	1845–1892	Railroads, iron, coal, construction
3rd wave	1892–1948	Electrical power, automobiles, chemicals, steel
4th wave	1948– ?	Automobiles, electronics, large home appliances, aerospace, pharmaceuticals, synthetic materials, composite materials

The concept of the economy being driven periodically by major investments in radical new technologies has both supporters and detractors. The phases of overbuilding and recession presumably include a stockpiling of technical advances ready to provide a new basis for investment after the recession has run its course.

The several directions of economic theory described briefly here, plus the great amount of economic research to which I have not referred, are all part of a broad effort to relate R&D resources and outputs to the economic environment in which technical activity is conducted. The economists who have been so productive in that field, Mansfield, Rosenberg, Nelson, et al., would probably agree that the results of their studies produce guidelines, not formulas. They provide an improved understanding of how economic factors influence, or are influenced by, the process of technical change. They are important tools to be used as such by policymakers in government and executives in industry.

We can learn from the relevant economics research that the economic values derived from the application of resources to R&D may follow a different pattern from that of technical advances. A fixed.

rate of technical effort may yield significant economic returns but a declining rate of significant technical advance.

Any technical advance can yield some economic benefit. The largest part of industrial research effort is intended to improve and strengthen existing products and businesses. Mansfield's work on productivity increases and R&D intensity implies that continued attention to R&D at any level yields increases in productivity commensurate with that level in that industry. Rosenberg's analysis calls attention to the economic returns that can accumulate undramatically as a modest technical advance diffuses from one industry to another.

When we consider the very substantial efforts of all companies to exploit technical advances or to develop market niches even for existing, state-of-the-art technology, it is clear that most technical activity is in fact more likely to produce economic returns than to create significant technical advances. This is certainly true for small and medium-size firms.

The exploitation of a technical advance requires considerable problem-solving and engineering design that is required to adapt the advance to a product or use. It may call for materials development, or ingenious circuitry or creative computer programming. While the total amount of such technical activity accounts for most of all R&D, it can be carried out in projects of reasonable size within the capacity of even small firms. Thus, the diffusion of technical advances throughout industry and the many specific actions to produce economic returns do not require massive or constantly increasing technical resources.

This point should not be taken as a contradiction of earlier statements. Constantly increasing technical resources seems to be needed in order to maintain a given rate of significant *technical* advances, and there are observations we can draw upon to relate the amount of technical resources or an increase in technical resources to any level of *economic* returns.

One of the real strengths of the U.S. economic and technical structure is its ability to develop economic benefits from very modest technical advances. The identification and exploitation of market niches, the ability to implement new technology through the addition of many modifications that fulfill economic needs or opportunities, all create economic growth without adding directly to continued technical advances of any magnitude. Indirectly, the opening of new markets provides financial incentives and resources for added R&D, and the greater effort does increase the probability of significant new advances.

While quantitative correlations do not exist between economic returns and technical advance, or cannot be formulated easily, there are likely to be different levels of economic benefits that attach to significant technical advances in contrast to "ordinary" technical progress. Presumably such significant advances would include those which are considered to have triggered the growth phases of the Kondratieff long waves.

There is much opportunity for intensive analytical economic research to explore the cumulative economic benefits from different forms of technical programs. The criteria for differentiating among technical advances would presumably include

1. Increase in understanding of a given field
2. Impact on other technical fields
3. R&D effort needed to produce the technical advance
4. Investment necessary to exploit the technical advance
5. Impact on range of industries

One key factor tying together concentration of technical resources, significant technical advances, and an economic return is investment. As mentioned earlier, there is a strong incentive for large technically based corporations to seek significant technical advances, since these require large resources to create and strong financial resources to exploit. Thus, a large corporation can "appropriate" much of the value from large complex R&D efforts because many potential competitors might not have the financial capabilities required for application. This is equally true of major cooperative R&D programs of a public nature. The more significant the technical advance resulting, the more likely that only a very large corporation or consortium can provide the investment necessary for its application.

The inherent difficulty in relating technical effort, technical progress, and economic benefit is that two separate systems are operative. Technical progress results from technical effort in a closed system within the control of those conducting R&D. It takes place within the technical enterprise. Economic benefits derived from technical progress occurs in the "open" system of the total economic environment and requires actions from sources not necessarily involved in generating technical progress. The state of the economy, availability of capital, interaction with public policies all affect the successful implementation of technical change.

PRESSURES ON THE TECHNICAL ENTERPRISE

Whatever the accuracy of economic theory, the belief that economic benefits flow from technical change is universally held. Even worries over technological unemployment resulting from automation are almost always overwhelmed by the desire to stimulate and apply increased technical activity as a basis for economic growth. Even in Europe, with its deep sensitivity to unemployment, there is no serious evidence of any modern-day Luddites.

Consequently, both private and public funding for R&D have gone up sharply in recent years. In almost all countries, there is support for increased university research, there is generally support for government R&D to strengthen the civilian sector, and there is often support for government funding of private sector R&D intended for corporate growth. This last item is always related to expected increase in employment and usually to potential for exports.

In addition to increases in direct funding for R&D, other actions serve to provide incentives to increase technical activity or to strengthen supporting activities. Tax reductions for incremental R&D costs are being used in the United States, and other tax mechanisms add incentives for investments in R&D and in its exploitation. Extending patent protection is another incentive. Actions to attract students for technical degrees serve to provide a reservoir of trained researchers for the future.

The common thread in all these public actions and more is the desire for economic growth, though there is a separate added pressure from the standpoint of national security. This places a considerable pressure on the technical enterprise to perform accordingly, and an obligation to provide government and the general public with the understanding that can guide effective actions.

The one reasonably probable result of these stimuli is that more technical activity will be undertaken. The very likely output of this is that technical advances will be made. It is possible with enough concentration of resources and breadth of effort that significant technical advances will occur in particular areas. Undoubtedly, this will provide the opportunity for economic growth.

Economic growth depends on factors that are not often explicitly accountable:

- Whether all other conditions in the economic environment are favorable
- Whether the stimuli for technical activity are random or purposeful
- Whether the technical enterprise can act most effectively

The first item is simply a cautionary reminder that technical advances make economic growth possible, not automatic. Willingness to invest is dependent upon judgments of domestic and international conditions that go far beyond the performance of the technical enterprise.

The second item is a reminder that, while technical activity in pursuit of added knowledge can be random, technical activity *intended* to create economic growth, including the generation of knowledge, is purposeful.

The third item refers to the one which can be addressed within the technical enterprise or, more precisely, by those in positions responsible for the planning and management of it. This is the obligation to improve R&D productivity itself, to generate the maximum amount of technical advances distributed among the many objectives placed upon the system.

The issue is really a question of balance and of understanding. Private expenditures on R&D are in balance with the economic system almost by definition; that is, the level and direction of industrial research is compatible with corporate resources and business plans for conversion to economic use. When the magnitude of industrial research is inadequate for the needs and opportunities for further economic growth, public and private mechanisms have been initiated with the involvement of the private sector. This presumably maintains the balance with the other productive resources necessary for converting technical advances to economic use.

Public expenditures or public policies to stimulate technical activity in order to produce economic growth may not be balanced. It is relatively simple for a government to appropriate money or to change a single tax regulation. It is not at all easy to create markets, provide incentives for adding manufacturing capabilities, or adjust foreign exchange rates, and it is impossible to do all these things simultaneously. Government carries out what is possible, not necessarily what is required. And it is always possible to stimulate technical activity.

At least, then, there should be understanding, so that (1) no great resentment should arise if anticipated results do not emerge and (2) a steady pressure can exist to stimulate related actions that would encourage economic growth based on the added technical activity. Much public policy in strengthening R&D is based on the belief in the sequence that technical activity produces technical advances, which lead to economic growth. This progression does not happen automatically. It calls for attention to the interactions within the technical enterprise and for the recognition that it is not a random process. Public policy can be strengthened and made more realistic when it takes into account the complexity and the potential within this system.

NOTES TO CHAPTER 10

1. Nathan Rosenberg, *Inside the Black Box* (New York: Cambridge University Press, 1982), pp. 280–291.
2. Jeffrey E. Garten, "America's Retreat from Protectionism," *New York Times*, June 16, 1985, p. F3.
3. Private communication.
4. "Now, the Japanese Challenge in Microprocessors," *Fortune*, July 8, 1985.
5. These networks are analyzed in a study by Carmela Haklisch titled "Technical Alliances in the Semiconductor Industry," Center for Science and Technology Policy, Graduate School of Business Administration, New York University, 1986.
6. See especially *The Theory of Economic Development* (Cambridge, Mass.: Harvard University Press, 1934) and *Capitalism, Socialism and Democracy* (New York: Harper & Brothers, 1942).
7. Robert Solow, "Technical Change and the Aggregate Production Function," *Review of Economics and Statistics* 39 (1957): 312.
8. Edward Denison, *The Source of Economic Growth in the United States* (New York: Committee for Economic Development, 1962).
9. E. Mansfield et al., *Technology Transfer, Productivity, and Economic Policy* (New York: W.W. Norton, 1982), pp. 117–131 and references cited there.
10. E. Mansfield et al., "Social and Private Rates of Return from Industrial Innovations," *Quarterly Journal of Economics* (1977).

11. See generally Rosenberg, *Perspectives on Technology* (New York: Cambridge University Press, 1976).

12. N.I. Rosenberg, "An Assessment of Approaches to the Study of Factors Affecting Economic Payoffs from Innovation: A State-of-the-Art Study," Stanford University, March 1975.

13. R.R. Nelson, M. Peck, and E. Kalachek, *Technology, Economic Growth, and Public Policy* (Washington, D.C.: The Brookings Institution, 1967).

14. R.R. Nelson, ed., *Government and Technical Progress: A Cross Industry Analysis* (New York: Pergamon Press, 1982).

15. J.A. Schumpeter, *Business Cycles: A Theoretical, Historical and Statistical Analysis of the Capitalist Process*, 2 vols (New York: McGraw-Hill, 1939).

16. D.B. Merrifield, "Strategic Planning and R&D Productivity in a Decade of Change," in H.I. Fusfeld and R.N. Langlois (eds.), *Understanding R&D Productivity* (New York: Pergamon Press, 1982).

17. R. Rothwell, "Innovating towards Prosperity," *Chelwood Review* (1982, Christmas Issue).

MANAGING THE TECHNICAL ENTERPRISE

11 TRANSFER PROCESSES WITHIN THE TECHNICAL ENTERPRISE

Every technical unit that conducts R&D generates inputs for the technical enterprise. The resulting flow of science and technology is available for use by every other technical unit, either through the public channels traditional for science or the numerous private channels more relevant for technology.

This simple characteristic provides three forms of benefit:

1. Every R&D performer has more knowledge available that may reduce costs or time in carrying out a program. This is a general benefit that can be increased through investment in technical liaison and in technical data networks.

2. A program can be planned with the intention of incorporating results from other programs. An example of this interaction is the use in the private sector of technical advances created through large public sector programs, as well as the reverse situation.

3. An objective that requires a technical effort may be approached in a manner that stimulates further effort throughout the technical enterprise, so that the initial funding of R&D produces a higher rate of technical activity contributing to that objective. This is particularly applicable when the effort required is in basic research and is implemented within the academic research sector.

This is not a philosophical discussion of technology diffusion, but a practical concern with how to convert a random process to a purposeful one. Two large categories of expenditure depend for their effectiveness on the transfer characteristics of the technical enterprise: large technical programs initiated by government, and the directed basic research at universities involved in the pursuit of a broader program.

INTERACTIONS BETWEEN FUNDING SOURCES AND ACADEMIC RESEARCH

The scientific base of the technical enterprise does not grow uniformly or at random. It is influenced by the allocation of funding. A feedback mechanism is built into the internal dynamics of any scientific field. New discoveries or developments generate further advances in knowledge or further applications. Thus, money drives technical activity, and technical opportunity in turn attracts money. Since today's objectives provide funding for preferred areas of basic research, which influence tomorrow's opportunities and objectives, the objectives and actions of one funding source with regard to academic research affect those of any other concerned with the same field.

Consider the progression of basic research activity in semiconductors. Basic theory and experimental observations on behavior of the solid state progressed during the 1930s and 1940s. Much of the research in semiconductors was conducted at universities, supported by government agencies after World War II, and government procurement paid for manufacture of semiconductors and transistors. These new components provided opportunities for civilian products in communications and computers. Because of its needs for continued scientific advances, industry then made private sector funds available for university research while greatly expanding basic research in its own laboratories. The resulting technical advances in integrated circuits led to advances in microprocessors, which in turn led to development of very high speed integrated circuits (VHSIC) and other electronic devices essential for defense. As a consequence the U.S. Department of Defense has plans to increase considerably the funding of university basic research in microelectronics.

One lesson inherent in this example is that the objective of one funding source contributes to those of another most easily when there is a common interest in areas of basic research. A second lesson is that this networking takes a long time. The semiconductor story has been going on for about 50 years, and no end is in sight. The transfer process is continuous, not immediate, and requires substantial investment as well as time. A similar network is growing in biotechnology, funded by billions of dollars from the National Institutes of Health over at least 20 years. Pressures from industry for basic research are growing as commercial applications expand.

Researchers in the universities may worry that requirements of funding sources distort the emphasis of university research from some "natural" balance. But it is hard to say what this balance would be in the absence of external funding influence.

Research areas grow or shrink in response to external pressures. Research interest is attracted to fields where new theories or new experimental techniques have become available. Growth of activity worldwide attracts new researchers. Public interest, whether because of national objectives, such as defense, or major economic concern, as with computers, focuses attention on the underlying basic scientific problems.

All institutions with an interest in a particular field of basic research benefit from the increased funding. Many examples exist where funding of basic research triggered much greater funding of the same or related areas, thereby expanding the field and yielding outputs exceeding that resulting from the original funds.

The university system acts as an amplifier for basic research within the technical enterprise. Above a certain threshold of funding, it seems, the concentration of effort may be great enough to create new technical opportunities and attract additional funds. This feedback provides opportunities to derive added benefits from the system through planning and selectivity, a process that is underway in current funding of research in microelectronics.

Considerable flexibility within academic research derives from the variety of funding sources involved. The very nature of basic research is such that the results can have application in areas other than the funded one, whether through theory or through new instrumentation.

Academic research thus not only links funding sources but transfers results within the university system. The transfer of new knowledge can be made more effective by conscious actions of the users. So long as this ability to act as a transfer mechanism in the area of basic research is preserved, the benefits for all funding sources are increased. Our concern should not be with the extent to which funding comes from government or industry, from the Defense Department or National Science Foundation, but with the inherent flexibility and transfer processes of the university system, and with the multiplicity of funding sources.

INTERACTION BETWEEN LARGE PUBLIC PROGRAMS AND COMMERCIAL TECHNOLOGY

The growth of the worldwide technical enterprise is stimulated by any increased R&D effort. A large public program contributes to the base of science and technology that can be drawn upon by private corporations for subsequent commercial use. Although that sequence is the usual starting point for public policy discussions, it is important to note that the reverse situation also holds; that is, a large industrial research effort strengthens the technical base for subsequent public sector programs.

GLOBAL PRINCIPLES

Public programs are of two kinds with respect to the generation and use of commercial technology:

1. Programs that do *not* have commercial application of the resulting technical advances as the principal public objective
2. Programs that *do*

The best example of the first category is defense, although other examples can be drawn from the health sciences and the initial space program. Not many examples of the second category can be called successful. The synthetic fuels program is one example in the United States. The Cooperative Automotive Research Program (CARP) in the late 1970s was launched but withdrawn shortly after.

PUBLIC PROGRAMS WITHOUT
COMMERCIAL OBJECTIVES

The effects of defense programs on the technical base of industry are always significant, often obvious, and usually complex. An excellent discussion relating U.S. defense research to civilian technology is contained in a recent book by Rothwell and Zegveld.[1] A powerful stimulus for technical change has been the use of government procurement. NASA has promoted high technology related to space and the NIH has been a major sponsor of R&D in the health sciences. NIH programs are not as large as the Defense Department programs, but they are specialized and intensive, and NIH has built a strong foundation in the health sciences over the decades.

Any increase in basic knowledge is available for industrial use internationally. The specific area of basic science pursued under a public program may not be precisely the take-off point for commercial developments, though there is a high probability that it will trigger research in related concepts. This is part of the amplification process. In any event, the easy transfer of basic knowledge only enlarges the scientific base.

The basic research conducted under the umbrella of a major public program provides some modest advantages for domestic industries, but not a unique advantage. Moreover, commercial development and conversion to products and processes is still a time-consuming process from concept to sale. The time required for the output of basic research to have an impact on industrial growth is considerable even when the work is done within corporate laboratories, and is greater when it is performed in universities or government laboratories.

We can expect the noncommercial objectives of major public programs to result in outputs of specific process know-how, new materials, products, and systems. The probability that any of this can be put directly to commercial use is extremely low. The probability that some of this can be modified, designed for economic manufacture, and applied to a narrow market niche is better, but still low. The probability that increased familiarity with research outputs will stimulate related industrial research that can eventually lead to commercial outputs is fairly good, as it is with any general advance within the technical enterprise, but that takes time.

New materials, products, and processes are forms of intellectual property, normally controlled by the public agency or the R&D contractor, depending on the policies of the government involved. However profitable conversion to commercial use may eventually turn out to be, it requires investment and commitment of personnel under the normal conditions of uncertainty that govern any technical development. Government policies regarding patent ownership, licensing for use, exclusivity, and financial terms can be an incentive or disincentive for private investment. These policies have to be flexible, since the effects will differ case by case.

Is this transfer process simpler, faster, and less costly when the R&D funded by the public program is performed by a corporation? The answer depends on the R&D subject, the product lines of the company, and the conditions under which the work is done.

The preferred situation to expedite commercial use of technology from public programs is to have the technical personnel responsible for the commercially oriented programs be the same ones who conduct the R&D under contract from the public programs. The second best situation is for the contract R&D to be performed by people in the same laboratory and location where the responsibility for commercial development exists. All other situations involving people at locations for contract R&D which are separate from the places where commercial development is conducted worsen the transfer process.

These general guidelines for *any* technology transfer process are well known to research managers. The most effective transfer takes place within the mind of a single individual, so that the knowledge and technical approaches experienced in one program become part of the technical base that is drawn upon to initiate new programs with different objectives or to suggest new approaches in the solution of some other problem.

In practice, the conditions under which contract R&D on public programs are conducted by corporations are not those which are most favorable for the transfer of resulting technology to commercial use. Obstacles exist that relate to personnel, to management, and to legal restrictions, at least in the United States.

The subject matter for noncommercial research, however relevant to the technical base of the company, is therefore not likely to be precisely the subject of greatest benefit to the business plans of the corporation. Even when all the *costs* are paid by public agencies, the program is not the highest priority use of the *time* of R&D personnel most experienced and competent in the planning and conduct of commercially oriented R&D. The public program is a diversion from the optimum R&D effort intended to provide commercial benefit.

Thus, a strong pressure exists within the corporation to focus the attention of people committed to the technical needs of corporate strategic plans on the company-funded R&D. Contract R&D is then conducted in a research group that is separate organizationally from the corporate-funded group, and very often separated physically. The defense R&D of General Electric is carried out at Valley Forge, Pennsylvania, several hundred miles from GE's major central research laboratories in Schenectady, New York. RCA concentrates defense R&D in Cherry Hill, New Jersey, an hour's drive from the corporate laboratory in Princeton.

Organizational and physical separation also serve to minimize legal difficulties for a corporation that conducts government contract R&D in addition to more traditional commercial business. Public programs include the payment of overhead charges that include general administrative costs of the corporation. Research equipment that is bought under a government contract should be used on that contract or at least on some government R&D contract. The simplest procedure to assure government auditors that the costs charged to government contracts were indeed incurred only because of those contracts is to place government work and the people involved in a location committed only to government contract R&D.

Much defense R&D conducted in industry is classified. The plans, approaches, and results cannot be communicated legally to other people in the corporation without security clearance. No convenient and economical procedure exists to embed a classified program within an organization committed to unclassified R&D. It is done on occasion with a single classified project, but any appreciable amount of classified research leads very quickly to a separate facility. If the company does not initiate this for financial reasons, the government may insist for security reasons.

In summary, public programs not intended for commercial objectives incorporate obstacles to transfer of technology even, or especially, when they are conducted under contract by industry. Government and university R&D represent delays of time and communication. Industry activities in contract R&D represent obstacles of physical separation of facilities and legal restrictions on classified programs.

Does this mean that large noncommercial public programs do not benefit industrial research and, therefore, industrial growth? Not at all, but the benefits are indirect, occur over a long period of time, and are diffused throughout industry. The billions of dollars spent on R&D by the Department of Defense, the Atomic Energy Commission, Department of Energy, NASA, and the National Institutes of Health since 1950 have built a technical foundation in the United States that is unmatched in the world. Government funding did not create products and processes directly, but it did create a reservoir of people and know-how that flowed into an industrial research structure that is also unmatched in the world. It strengthened the university system and expanded the pool of people attracted to science and engineering.

This raises an interesting question. If the important commercial benefits of noncommercial public programs are indirect, why not simply spend more money on R&D? If we have political concerns about defense, environmental questions about energy, economic challenges to space, then why not dispense those same billions of dollars, or at least those billions not needed for the actual building of hardware, to universities and research institutes, for example? Isn't any expenditure of a billion dollars in R&D as good as any other if the principal benefit expected is the indirect strengthening of commercial technology?

The answer is no, but the reasoning may not be immediately obvious. Besides the legitimate national objectives to be achieved in defense, energy, and space, a mission provides discipline for the organization and coherence of the many technical activities needed to accomplish technical progress.

Science and technology provide a wide scope for activity. Some progress occurs each time a serious and well-planned R&D project is completed, whether by an individual or a group. Progress is accelerated, and significant technical advances can occur, when activities are simultaneously pursued in related areas that reinforce each other, particularly when theoretical work and experimental programs are related. A technical mission provides a focus for R&D, often attracting both scientific and technological efforts that emphasize whatever mutually reinforces and, for the moment at least, puts aside whatever does not.

Consider the development of the nuclear reactor. Decades of advances in nuclear physics had provided the theoretical base for the concept of a nuclear reactor. No comparable base of knowledge existed in the materials field that would permit its construction. Presumably one course of action would have been to initiate massive funding of basic materials research to broaden our understanding. Many interests would be pursued, and ultimately we would be in a better position to construct a reactor.

The technical mission, of course, was to build and operate a nuclear reactor. Out of all the areas of basic research in materials that were possible, research effort concentrated on those questions about the structure of solids that related to the construction of a reactor. What happens in a solid when neutrons bombard it? Are they absorbed? By what atomic or nuclear mechanisms? What is emitted and by what processes? Are structural defects created? How are those various processes affected by the atomic structure?

The mission formed a marriage of nuclear physics and materials science. It advanced our basic knowledge of both. If the same amount of money had been spent without the mission, we would have had small advances in many directions but low probability of significant advances in any.

The increased probability of technical advances from the discipline of a mission objective has been demonstrated with jet engines and supersonic aircraft satellite launches, putting men on the moon, and for laser-guided missiles as well as a broad-based attack on cancer.

A mission that requires major technical change serves to concentrate resources and to provide purposeful linkages among the vast number of technical activities that can be pursued. Hence, while noncommercial public programs benefit industrial growth indirectly, they do so more effectively than a simple increase in R&D funding. An across-the-board increase in R&D activity strengthens the technical infrastructure. A major public program can result in significant technical advances.

PUBLIC PROGRAMS WITH COMMERCIAL OBJECTIVES

The objective of some government programs is not technology, but economic growth. Normally, this translates into jobs and exports, which are based on the manufacture of goods and the delivery of services. The critical aspect is that the mechanisms by which technology is converted into products, processes, and services reside in the private sector, within industrial research.

Several detailed studies have appeared recently concerning government programs which affect industrial technology and growth. One thorough background on U.S. industries is the work edited by R.R. Nelson.[2] Another analysis, with data from European experiences, is the previously cited work of Rothwell and Zegveld.[3]

Publicly sponsored commercial R&D appears to be either generic, with applications in more than one industry, or industry specific. Support for generic science and technology has been the most frequent and most acceptable form of public program for strengthening commercial technology. Such programs do not involve government in decisions with regard to marketing, conversion of technology, or investments. A clean line of demarcation exists between the technical participation of government and the operational decision-making of the private sector.

Public support for generic programs has been highly productive in the United States for agriculture and, in its formative days, for the aircraft industry through NACA. Two aspects of these cases seem relevant generally. First, the nature of the government R&D activity meshed well with the private sector needs and structure. Second, the users participated in the planning of the programs and then continued to interact with the researchers during conduct of the R&D.

The activities of NACA, discussed by Mowery and Rosenberg,[4] provide a good example of government-industry interaction. While NACA's was clearly a public program conducted in public facilities, it appears that responsible technical personnel from the aircraft companies took an active role in discussing technical needs of the new industry and in maintaining close contact with the ongoing technical activity. The government developed major experimental facilities at a time when the individual companies did not possess the resources. Of great importance was the consistent and accepted public policy aimed at encouraging the emergence of a strong aviation industry. This example of workable government-industry cooperation occurred under greatly different political and economic conditions than exist today. Since we are still striving to create effective mechanisms for technical advance, there is value in reexamining the lessons found in NACA.

One example that did not work out well was that of the Cooperative Automotive Research Program (CARP) initiated in the late

1970s. The motivation was straightforward, namely, that the U.S. automobile industry was losing a substantial share of the domestic market. Presumably this trend might be reversed by technical advances in performance and perhaps in lower manufacturing costs.

While there were many things government *could* do to encourage more R&D in the automotive industry, that industry already maintained two of the largest research laboratories in the world, and all manufacturers had large and sophisticated engineering staffs—a situation quite different than the case of a young aircraft industry or a disaggregated agricultural industry with little research. The initiative for CARP came from government, which is not unreasonable, but is not preferable to industry initiative. More damaging was the industry's perception that the program was being imposed upon it, rather than emerging from it. Although the responsible government officials did attempt to develop a dialogue with industry and to solicit program recommendations, there was little internal pressure from the industry for a program of cooperative research.

Of importance was sensitivity of the industry, particularly the largest corporation, General Motors, to the potential of antitrust violations. Despite government initiatives in organizing the planning and discussions, an industry with three dominant corporations, one of which towered over all U.S. industry, was not enthusiastic about presenting a possible target for future antitrust changes.

Perhaps more important was that the leading corporations, certainly GM and Ford, had large laboratories and great financial resources. There may well have been good reasons for not pursuing certain areas of R&D. The automobile industry is a complex mix of technology, styling, marketing, and pricing. All affect the competitive position of the individual company and of the domestic automotive industry generally. The hypothesis that a government-initiated cooperative research program would make a major difference in the competitive status of U.S. automakers was clearly not accepted by the companies.

This is an example of a difficulty that has occurred more often with public programs to develop specific products or processes. It may be a question of picking winners, and the U.S. government does not have a record of success.

Government's support for the SST, for example, is in sharp contrast to the NACA case, an interesting point since both are in the same industry.[5] Whereas NACA emphasized testing and research, development of the SST was to specifications of the Federal Aviation Agency (FAA) and was intended to result in a procurement award. It was an attempt to use the military approach of development and procurement for a major system which would have to meet commercial economic criteria. Simply describing the program in these terms is a statement of the problem. Government development and procurement works well when the government is the customer. The government's attempt to stimulate a major entry into civilian markets through the same procedures resulted in an uneconomic product that the airlines could not buy and operate in a competitive market. The program was stopped by Congress in 1971, leaving the commercial SST field to the Concorde.

The lessons from such experiences in the United States seem to boil down to one simple guideline: Public programs intended to provide commercial technology must be based upon the plans, the needs, and the involvement of the companies which will absorb and exploit that technology. This is painfully obvious when commercial decisions are required. It is even valid when basic research or generic technologies are the substance of the program, since the selection of projects and the transfer of results are most effective with active cooperation of the industrial research community most affected.

SPECIAL CASE: DEFENSE BUSINESS AND CORPORATE MERGERS

Discussions of how public programs involving R&D affect commercial technology most often focus on defense R&D for two reasons. First, the amount of money for these programs is very large, and second, defense R&D includes considerable effort in advanced technologies (materials, microelectronics, computers, telecommunications, remote controls) and much advanced work in established technologies (manufacturing processes, electromechanical systems, lubricants, catalysis, construction techniques). In short, the incentives of big money and major technical advances are attractions for industry attention.

One approach to the benefits of defense funding is *not* to develop and transfer technologies for commercial exploitation, but simply to operate a large defense business. The size of market is huge and growing, and the market is perhaps more diversified than it may appear at first glance.

Technology transfer in the defense business is direct. That is, the R&D is paid for by the user (the government agency), who establishes the specifications for the final product. The corporation conducting the R&D as part of a defense business would normally wish to be a supplier of the final product. Programs of R&D are conducted with specific applications in mind. Even though the R&D may be paid for by the government, and is a "product" that is sold, the large financial returns come from production. The company may be willing to conduct R&D, at little profit or a modest loss, in order to generate a technical advantage for later production. In this sense, defense R&D by a defense supplier follows guidelines similar to those in more traditional industrial research.

In brief, there is a definable defense industry within which R&D is conducted for use in that industry, without necessarily giving too much thought to transfer and use in civilian markets. Nevertheless, significant technical advances can result because of the difficult performance specifications and the massive resources devoted to this activity. This pushes forward frontiers in many areas of technology, provides sophisticated know-how for a substantial number of scientists and engineers, and thereby creates a reservoir of knowledge and skills that can strengthen the overall industrial base indirectly over time. As a technology advances, breakthroughs at the previous frontier eventually become declassified and available to the public.

Consider defense electronics, one of the most intensively pursued high-technology activities. Multibillion dollar programs are funded through defense R&D and procurement in such areas as

- Very high speed integrated circuits (VHSIC)
- Supercomputers
- Artificial intelligence
- Gallium arsenide and other electronic materials
- Fiber optics
- Lasers
- Robotics
- Imaging technology and simulators

These are all at the core of the technologies affecting much of industrial growth in high technology, and much of the technical change sweeping the mature industries. Putting aside for the moment questions of civilian use, it is interesting to note the sheer magnitude of these technical activities within the defense industry.

The following data for spending on defense electronics in fiscal year 1985 were assembled by the Electronics Industries Association and were reported along with a broad analysis of the defense electronics industry.[6]

Research, development, testing, and evaluation (RDTE)	$14.7 billion
Procurement	$31.4
Operations and maintenance	$3.8
Total	$51.9 billion

These numbers can be compared with the total Defense Department budget of $282 billion in fiscal year 1985 and total RDTE expenditures of $30.5 billion. The defense electronics expenditures are roughly 15 percent of defense expenditures, and constitute about 48 percent of all defense technical activity. More relevant to the technical enterprise is that company-funded R&D for 1984 in the industry sectors identified as electrical, electronics, computers, office equipment, peripherals, software, semiconductors, and telecommunications all add up to about $16 billion.[7] Considering the broad interests of the companies, defense electronics R&D matches private expenditures in the technical areas concerned.

Here then is a major portion of the technical enterprise devoted to a particular group of high technologies, but focused on the needs of one category of user—military establishments. A sobering note is that the development of advanced military items has become a major export business. There are severe restrictions on what electronic innovations can be exported, but considerable pressure exists from such classical sources as competition and balance of trade.

A world market in military areas and systems of about $35 billion in 1984 was supplied as follows:[8]

USSR	$9.4 billion	27%
Western Europe	8.1	23
United States	7.7	22
Other	9.8	28
Total	$35.0 billion	100%

Since Europe spends 4 percent of its GNP on armaments compared to the 7 percent of GNP spent by the United States, there is more pressure on European companies to seek these exports. France is the most vigorous exporter, accounting for half of Europe's arms exports. Much of the dollar value of these exports is in hardware such as tanks, planes, and weapons. However, the electronic content of these items is substantial, and increasing emphasis is on advances in electronic guidance and communications. Competition in these markets is a strong incentive for R&D. It is an important factor in current European discussion about joining the SDI program.

The defense business is thus a major industry sector which puts great resources behind the pressure for technical change necessary for that industry. It advances the technical enterprise generally, but with restrictions. It is a particular contributor to advances in high technology. The dissemination and use of these advances for economic growth is a challenge and temptation, but remains largely indirect.

Commercial applicability of defense technology and the fact that defense electronics is a major industry in its own right have led to some large corporate mergers or acquisitions. Two of the largest were the Sperry Rand Corporation, formed by the merger of the Sperry Gyroscope Company and Remington Rand, and Rockwell International, which combined North American Aviation with the Rockwell Company. More recently, we have witnessed the $5 billion purchase of Hughes Aircraft by General Motors.

Such corporate marriages are based on economic considerations. The defense partner was evaluated on traditional factors of investment, profitability, market share, and growth. Joining of a major defense supplier with a major civilian manufacturer provides diversification for the combination. In each of these cases, deliberate attention was given to the potential economic benefits for applying the high-technology outputs of the defense partner to the businesses and skills of the commercial partner. Roger Smith, chairman of GM, stated that "Hughes is the key to the 21st century for GM; it is exactly what General Motors needs to get where GM wants to be."[9]

This approach is hardly restricted to the United States. The purchase of a controlling interest in Dornier by Daimler-Benz is part of a diversification program into high-technology industries. At the same time "Daimler is looking for a double pay-off. Dormier's research in aerospace, electronics and such new materials as carbon fibers and industrial ceramics will help Daimler develop advanced technology for cars, trucks, and buses."[10]

How successful have these "mixed marriages" been as business entities and as technology transfer mechanisms? Perhaps the only real clue is that the older marriages have survived and presumably prospered by maintaining the defense and civilian portions as separate businesses. The benefits of diversification in two different parts of the economy and the options this provides for investments appear to be real and well established. Technology transfer, on the other hand, has the same opportunities and difficulties as transfer from any division of a large corporation to any other, with two additional obstacles:

1. Results of classified defense R&D cannot be transmitted.

2. Results of defense and civilian R&D tend to breed two cultures whose populations are not easily interchangeable. Defense emphasizes performance, not markets or cost, though regard for the latter may be changing. Traditional industrial research emphasizes technical advance compatible with an economic system but does not operate through competitive proposals and contract management.

For example, senior technical executives at Sperry and Rockwell, past and present, have made efforts to communicate, to develop liaison, and to set up ad hoc R&D cooperation, when there is a need and opportunity, between defense and civilian technical activities within their corporations. But there are no examples where a defense technology has been the basis for significant economic benefits in the civilian part of the company, certainly not at a level that would justify corporate mergers. Despite the differences between defense contract research and traditional industrial research, some transfers of personnel do occur, and they can produce definite benefits.

Can GM derive anticipated high-technology applications from its marriage with Hughes? Senior technical executives at GM believe so, partly because they are knowledgeable about the earlier examples, partly because they are familiar with the process of technical change. Their experiences with the entrepreneurial group of small companies in robotics and artificial intelligence are encouraging with regard to the corporation's absorbing technical advances.

An important criterion for successful technology transfer will be the expectations of top management. The size of Hughes suggests that its own profitability and business operations will always take first priority for its management and technical staff. Modernizing the automotive industry is a valuable technical goal, but it is unlikely to be pursued by Hughes' technical personnel as a separate objective. Since Hughes is a contract R&D organization, this suggests that programs of interest for automotive applications can be provided by Hughes through the use of those same mechanisms.

The point is that the transfer process from defense R&D to civilian use requires substantial effort to produce substantial returns. The articles in magazines and newspapers about corporate marriages between such partners often imply that one simply moves a company active on high-technology defense frontiers up against an established industrial concern and—boom!—a transfer of technology occurs. The potential is there, but it was there when the two were separate. The shared corporate umbrella may provide an illusion of common interests, but it does not in any way lessen the obligation for constructive planning, creative relationships, hard work, and time.

RECENT LARGE PUBLIC PROGRAMS

Efforts to concentrate technical resources European-wide have proceeded at different times and in different areas since the early 1970s, driven by a sense of falling behind the United States and Japan technologically. The larger U.S. markets, the cohesion and homogeneity of Japan, each provided the base for large industrial research organizations, while the U.S. university research effort mushroomed with heavy government support. All of this seemed to drive home the fragmented markets and the relatively smaller research organizations of European corporations. The several exceptions such as Philips, Shell, and Siemens did not change the perception.

In the scientific field, with no economic obstacles to cooperation, the European Science Foundation was set up in the early 1970s to facilitate cooperative activities. The ministers from 21 European countries agreed in September 1984 to develop scientific cooperation and encourage networks for scientific research within Europe. The ministers stated that such networks and greater mobility for researchers were necessary to give Europe's scientists "the sense of cohesion needed to remain competitive at both the scientific and technological level with the United States and Japan." French Prime Minister Laurent Fabius called it "uniting to survive."[11]

Other collective research efforts in Europe were discussed in earlier chapters. At this early stage, clearly the most successful is ESPRIT in terms of a working plan, participation of major firms, size of commitments, and initial R&D efforts. The so-called precompetitive activity is more applied research than basic. An important by-product in its early years has been the increased activity in joint ventures and cooperative research by companies involved in ESPRIT, due in part to the greater familiarity in working together within that program. The Commission of the European Communities, which provides the structure and half of the funding for ESPRIT, has been encouraged to set up a new program for telecommunications research (RACE) and to begin planning for some form of collective activity in biotechnology.

This activity is accelerating, and must be considered together with related activities, such as Franco-German cooperation in electronics and computers, adding government funds to R&D programs of industry. Edith Cresson, Minister of Industry in France, said that "French-German cooperation should be the line for promoting an increase in European cooperation as a whole."[12]

The ferment stirred by such activities is real, certainly among the national governments and international bodies, and somewhat within the companies. However, the largest program ESPRIT at roughly $1.25 billion over five years is only a fraction of industrial research effort in that field within Europe, slightly more than the annual R&D budget of Philips alone. The companies are certainly not looking to ESPRIT as a replacement for their technical activity. Nevertheless, it does seem to have triggered new thinking about further actions that could be taken within Europe across national boundaries to strengthen the technical base of individual corporations, and of Europe as a whole with respect to the United States and Japan.

Now consider the effect on those psychological stirrings within Europe of the U.S. proposals for a Strategic Defense Initiative (SDI), the "Star Wars" program. Here was a plan that would allocate $26 billion on essentially R&D activities over a five-year period: 20 times larger than ESPRIT. It would concentrate the efforts of over 7,500 scientists and engineers now working in related areas, growing to perhaps five to ten times that number in the later years of the program. The technical activities outlined by the Office of Innovative Science and Technology (IST) within SDI would include "reliable and durable microelectronics . . . capable of operating reliably for as long as seven years." The program requires supercomputers that far exceed the capabilities of today's fastest Crays and Cybers, as well as software that contains 10 million lines of error-free programming code. Much emphasis goes to advanced laser generation and control, optical sensors, new materials that would include organic molecule memories for computers, and a host of scientific and technological pursuits.[13]

The bulk of the R&D would be performed under contract at industrial laboratories. However, the university community is being attracted by a number of basic research objectives to be funded by a planned $100 million in 1986 and targeted for more as the program unfolds. Six research consortiums composed of universities, industry, and government laboratories have been initiated, with others being planned. These include

- $19 million over four years on nonnuclear space power to Auburn, State University of New York at Buffalo, Polytechnic University of New York, Texas Tech, and University of Texas at Arlington
- $9 million over three years on optical computing to Carnegie-Mellon University, Georgia Tech, Stanford, Lincoln Laboratories (MIT), Battelle Columbus, University of Alabama (Huntsville), and the U.S. Naval Ocean Systems Center
- $4 million over three years on improved electronic systems to University of California (Berkeley), Stanford, Purdue, University of Florida (Gainesville), and University of Southern California
- $2.5 million over three years on electronic systems and improved energy systems to the State University of New York (Buffalo), Naval Research Laboratory, and GE
- $15 million over three years on composite materials to MIT, Pennsylvania State, Colorado School of Mines, Johns Hopkins, Texas A&M, Brown, Rensselaer Polytechnic Institute, Drexel, SRI International, Naval Research Laboratories, plus several companies, including Martin Marietta, Fiber Materials Aeronautical Research Corporation (Princeton) and Ultrasystems

- $12.5 million over three years on spacecraft radiation, exhausts, electromagnetic waves, and particle beams to Johns Hopkins, Universities of Arizona, Maryland, Michigan, Iowa, Kansas and California (Berkeley), plus Utah State, University of California at Los Angeles, Massachusetts Institute of Technology, Stanford, New York University, Naval Research Laboratory, and the Air Force Geophysics Laboratory

A brief glance at the subjects and participants indicates the stimulus to further work through the interactions within the technical enterprise. The coupling of university research with some government and industry involvement strengthens the total base of the effort. Each project is modest, but the attraction of funds and interest can accelerate such efforts, particularly with the high probability of sharply increased funding ahead.

Technical and political arguments about both the feasibility of the SDI objectives and/or the implications for military strategy and U.S.-Soviet agreements aside, government spending of $26 billion in five years of R&D would have a great impact on many advanced technologies. The potential for a $2 trillion program to construct a full-scale space shield against nuclear missiles is many decades and many political debates away. The $26 billion for R&D, or less as Congress wills, is here and moving.

Unquestionably, these activities will raise the technical level of U.S. universities, of new graduates from those universities in science and engineering, and of the knowledge and know-how in the technical areas under development. Much of this will be disseminated worldwide through the technical enterprise, and will add to the reservoir of science and technology. Nevertheless, benefits will flow more quickly and effectively to U.S.-based corporations which will hire the new graduates and can maintain convenient liaison with the research activity.

What reactions have occurred in Europe to this potential? Most discussion, proposals and plans in the summer of 1985 had little to do with the political judgments on the broad SDI proposal, and very much to do with the implications for industrial growth and the competitive status of European corporations in the next decade.

One option is to develop relations between European corporations and the SDI program. A general U.S. invitation to this effect came first from President Reagan, followed by more direct "selling" by top representatives of DOD and the SDI program in the spring of

1985. European governments held back because of political uncertainties regarding long-term strategies and short-term considerations of U.S.-Soviet arms talks. European companies engaged in more specific discussions. Logica, a British software company, the French arms and electronics company, Matra, and the West German aerospace companies, Dornier and Messerschmitt-Bolkow-Blohm, among many others, expressed interest.[14]

Discussions in May and June appeared to be the prelude for widescale European linkages, at least through private corporations, with SDI. West Germany, Italy, and the United Kingdom indicated general interest, so that corporations in those countries were encouraged to explore the possibilities. As expressed by a British official on a NATO advisory group, David Hobbs, "Six months ago, the focus of the debate was strategy. Now, the debate is: What about the technological spin-offs? Can we afford not to participate?"[15]

Against this increasing momentum was the French proposal for a civilian-created cooperative European effort, the European Research Coordinating Agency (Eureka). French high officials say that much thought and planning has been underway on this for some time.[16] However, President Mitterrand's proposal for Eureka in April 1985 inevitably was treated as a quick reaction to the SDI proposition. It was hardly as attractive, since it had no specific program plan, certainly no promise of substantial funds, and presented formidable problems of political control and reaching consensus among governments and companies.

Perhaps surprisingly, then, a shift in attitude among both the governments and companies appeared to occur in mid-June of 1985 and to take form in the summer. The shift was not necessarily away from SDI, but it did represent first, a growing skepticism about precisely what was meant by "participation," and second, a growing interest in the possibilities for modifying and strengthening something like the Eureka proposal.

It is of great interest to examine the several lines of reasoning behind this shift. They relate to the themes of this book.

Public debate about SDI made it clear, as mentioned, that the immediate European concern was with commercial technology. Logically, then, two questions arose:

1. If a European company participates in SDI, what restrictions will there be on the subsequent use of the know-how for commercial applications?

2. If the pressure to generate technical advances for continuing corporate growth requires resources beyond the capacity of a single company, why not develop cooperative mechanisms to approach this directly rather than indirectly through SDI?

The original Eureka plan appeared too vague, quite possibly because France believed it necessary to come forward prematurely to develop some alternative to SDI. To many people in industry, and some in government, a general public program to stimulate R&D had little chance to make the kind of significant technical progress that would strengthen any commercial base. One commentary on Eureka in June 1985 was that "Bonn and London don't want yet another costly bureaucratic operation in Europe. One German Foreign Ministry official says he'd prefer clear, tightly run projects such as a European drive to develop supersmart observation satellites." [17]

This skepticism about specific contributions from broad publicly sponsored programs for technical cooperation was well founded, unless the object was simply to strengthen the technical infrastructure. But now one example existed which attempted to focus on well-planned projects. ESPRIT, while still very young and carrying a burden of international bureaucracy, had at least succeeded in having corporations work together to plan constructive technical programs that resulted in specific R&D projects. Could the too general Eureka proposal be converted to a project-oriented cooperative effort with the same objective of broadening the commercial technical base of European companies?

With the ESPRIT example in mind, imperfect though it might be, GE of Britain, Thomson of France, Siemens of the Federal Republic of Germany, and Philips of the Netherlands established an informal group to identify suitable cooperative programs. By July 1985, the group had examined cellular radio, high-speed rail systems, computer-controlled highways, and specialized energy programs. [18]

An additional concern has been the possible impact of SDI on the exports of defense-related equipment which, as noted earlier, amounted to $8 billion for Western Europe, about $4 billion for France. The SDI program would yield direct advantages to U.S. manufacturers for any systems incorporating advanced defense electronics. This has been addressed indirectly by the Independent European Program Group (IEPG), established to coordinate procurement policies of the European members of NATO. Acting within the

IEPG, a new program for cooperative defense research has been planned. Five areas of advanced technology have been chosen, including microelectronics and image processing.[19]

Underlying much of European coolness with SDI is the increasing concern with the U.S. policies on export controls. The possibility that vital components supplied by the United States could be shut off has been perceived as a threat to telecommunications systems in Europe. In discussion with senior technical officers of several British electronics firms and executives of the Alvey Directorate, I was told repeatedly that independence from U.S. supplies because of export controls was a major objective of R&D planning both at the companies and in the collective R&D activities.

This concern comes to the forefront with regard to SDI. Restrictions on export controls might conceivably prohibit European participants from using the outputs of SDI contracts for commercial use, a principal objective of participation. This sense of frustration has resulted in proposals by the North Atlantic Assembly to set up a new Technology Transfer Agency that would recommend mechanisms to satisfy both European and U.S. interests. The direct tie between this concern and other cooperative activity in Europe is discussed in a report to that Assembly by Lother Ibrüggen, member of the West German parliament, which sums up a great deal of the prevailing philosophy as follows: "If the alliance countries want to cooperate with the U.S., they must first learn to cooperate among themselves. If Europe could coordinate its own research and development activities better, it could compete better with the U.S., and then it could cooperate better."[20]

The next step in European collective activity is not clear. Individual corporations may very well participate in SDI. Other collective activities may be set up within the European Communities. A modified form of Eureka may be established outside the EC. All of these actions may be taken.

It does seem clear that the European intent to concentrate technical resources will lead to new forms of collective R&D activity. It seems likely that the public programs will emphasize specific missions rather than general increases in scientific activity, so that significant technical advances can occur. Finally, it is very likely that the focus on economic growth will result in the leadership within those collective programs being exercised by the industrial research community with the encouragement and blessing of all governments involved, conservative and socialist. The internal dynamics of the technical enterprise are becoming better utilized, if not completely understood or articulated.

NOTES TO CHAPTER 11

1. R. Rothwell and W. Zegveld, *Reindustrialization and Technology* (New York: Longman, 1985), pp. 139–148.
2. R.R. Nelson (ed.), *Government and Technical Progress: A Cross-Industry Analysis* (New York: Pergamon Press, 1982): see especially Nelson's own analysis on pp. 451–482.
3. Reindustrialization and Technology.
4. In Nelson (ed.), *Government and Technical Progress.*
5. Ibid.
6. Oppenheimer & Company, "Industry Review: Defense Electronics Industry," Report No. 85–907, July 8, 1985.
7. *Business Week*, "R&D Scoreboard," July 8, 1985.
8. *Fortune*, "Europe's Arms Exporters Challenge the Superpowers," August 5, 1985.
9. Quoted in *Wall Street Journal*, "High Tech Drive: GM May Set New Trend," June 6, 1985, p. 14.
10. *Business Week*, "From Autobahn to Aerospace," May 20, 1985, p. 80.
11. Quoted in *Science 225* (September 28, 1985): 1455.
12. *Science & Government Report*, January 15, 1985, p. 5.
13. *Physics Today*, July 1985, p. 55; *Business Week*, April 8, 1985, p. 77.
14. *Wall Street Journal*, "U.S. Tries End Run on Star Wars," June 10, 1985, p. 27.
15. *Wall Street Journal*, "West European Firms Seek Role in 'Star Wars,' Mindful of Widening Technology Gap with U.S.," May 23, 1985, p. 34.
16. Interviews.
17. *Wall Street Journal*, "Star Wars Plan, French Eureka Offer Closed Europe's Efforts on Cooperation," June 4, 1985.
18. Peter Marsh, "Star Wars Invitation Gets Cold Response in Europe," *Science & Government Report 15*, no. 12 (July 15, 1985).
19. *Science*, "Europe Agrees on Joint Defense Research Program," *229* (July 5, 1985): 36.
20. *Science*, "Europeans Seek Technology Transfer Agency," *226* (November 30, 1984): 1057.

12 FINE TUNING THE TECHNICAL ENTERPRISE
Financing

One start-up company that made good is National Semiconductor Corporation, which today has $1.5 billion in sales annually. Its founders, both physicists, were my friends Bernard Rothlein, president, and Edward Clarke, chief operating officer, who had led semiconductor development and manufacturing in Norwalk, Connecticut, for the Sperry Rand Corporation.[1] Their decision to produce low-cost, high-quality silicon transistors themselves led them to leave in 1959, after they had obtained financing, and set up manufacturing in Danbury, Connecticut.

A casual dialogue about starting a company had gone on between the founders for some time. The decision to *really* do it was made over hot pastrami sandwiches in a New York deli. A search for financing began. Following numerous discussions with venture capitalists, Rothlein and Clarke raised the initial funds. These had to be replenished at a later date by Peter Sprague, currently chairman of the company. By that time, the company had grown but had begun to have troubles in the marketplace; new operating management was brought in and both Rothlein and Clarke left, wealthier and undoubtedly wiser about the growth process of start-up companies.

The personal energy that goes into start-up companies was brought out in a conversation about six months after the launching of National Semiconductor. My friends were properly proud that in a

245

short six months they had a working manufacturing line producing transistors. They felt that no large corporation could have done as much so quickly with minimal resources. It turned out that they were putting in 70- to 80-hour weeks, plus "spare time" at home. Their commitment to their creation and their future was far greater than the efforts typical of commitment to an employer.

Organized industrial research, conducted within large corporations, has become the major force for continuing the growth of the technical enterprise. However, the benefits from technical advances can reach society through the activity of smaller companies, often started to exploit the opportunities presented by a particular new technology or scientific concept.

Technical entrepreneurship is a form of fine-tuning within the private enterprise system. A large corporation will not pursue a given technology through all possible applications since that may not be the best use of its capital and organization. The corporation may not be able to move fast enough even when a market is identified. By focusing on strategic growth, it may simply overlook an opportunity that differs appreciably from that growth plan.

Hence, an economy composed only of large corporations would account for most of the technical advances, all major manufacturing and utilities, and a very substantial part of the GNP. It would obviously miss many services, many consumer items, and a certain number of those areas of large commercial growth which can start from a small base without massive investment or organization. In some centrally controlled societies, those opportunities not provided by the large organizations do remain lost. In the market-driven economies, and particularly ours, these niches are filled by the entrepreneur raising funds from a variety of sources. Interestingly, one of those sources is the large corporation. These mechanisms are used in our form of society to fine-tune the process of technical change, so that the introduction of new technology into the economy is accomplished by a combination of large corporations and entrepreneurial units.

Clearly, the driving force in a new business is the individual. There are many individuals of great creativity, initiative, and drive who work well within the structured systems of large corporations. Great resources are made available to them, and they have used those resources well to build the massive industrial research base emerging in this century.

Other individuals prefer the freedom of individual decision-making without the structural procedures of large corporations. This freedom and flexibility compensate for limited funds and other resources. The incentive is the prospect of substantial financial gain from a successful enterprise, but the risk is high. The financial rewards actually appear to be less crucial motivation to most entrepreneurs than the desire for personal control and freedom of action. This is the basic psychological difference between owner and manager.

ROLE OF START-UP COMPANIES

Within the technical enterprise, start-up companies act to convert technical advances to commercial applications and link them to the marketplace quickly. Thus, they carry out a rapid diffusion of technology, thereby providing feedback about manufacturing processes, market demands, and technical performance. By identifying technical obstacles as well as real growth opportunities that can justify major investments, this feedback quickens the pace of technical change.

Many people believe that small companies are more innovative. Whether or not this is true depends on what we mean by "innovative" and on other facts. It is important to consider this point carefully, since much public policy designed to encourage or strengthen economic growth attempts to do so by emphasizing innovation, and in turn small business.

Policy discussions on this subject include three quite separate words—*small, new,* and *technical*—in referring to activity for bringing technical advances into commercial use through entrepreneurial activity. There is a tendency to use the three words interchangeably in connection with such activity. Almost all new companies are indeed small, but only a fraction of them have to do with commercializing a new technology. Most small companies are not new. Nevertheless, advocates of government support for all small business sometimes justify the legislative and executive actions they propose by claiming it stimulates economic growth because of the rapid application of technical advances.

On purely economic grounds, no such rationalization is needed to justify encouragement of small business. Solid data exists to demonstrate the growth of employment in the United States within the small business sector over the past decade. Small business has served to fine-tune the whole economy in identifying and providing services and products for both consumer and industry that large corporations overlooked or could not produce at acceptable cost.

If encouragement of small business is a good thing in general, what harm is there if policymakers are a bit careless in relating these actions to a desire to encourage innovation? One answer is that the desire to promote the application of new technologies through start-up business is diffused and made more costly by incentives for *all* small business, including those in existence which are far removed from any connection with technology.

A more important answer is that general support of small business to encourage innovation, based on imprecise definition and generally accepted relationships, can lead to policies which can weaken the process of technical change.

For example, a start-up technically based company, after creating a market niche with a new technological application, is very often bought by a large corporation. People who accept too literally the claim that "small businesses are more innovative" can argue that such takeovers decrease the economy's capacity for innovation and they might support policies to make such takeovers more difficult. But since the prospect of purchase by a large corporation is an incentive to the investors in and founders of start-up companies, preventing takeovers might seriously restrict their formation.

In a similar vein, proposals to modify antitrust laws in order to permit concentration of R&D effort arise from difficulties that large corporations may have in maintaining an adequate technical base in complex, fast-changing fields. Without passing judgment on the need for such modifications, those policies might be opposed on the grounds that such collective action causes small business to fall further behind, thus decreasing our innovative capacity.

The fact, of course, is that start-up technically oriented companies, usually small, and major corporations with large technical staffs *both* contribute to the process of technical change, *both* provide opportunities for creativity, and *both* produce economic growth. *Innovation* in this context covers the range of activities involved in the generation and application of science and technology, including the introduction of commercially acceptable products, processes, and services. Large corporations engage in all of these, but have a unique role in concentrating technical resources to generate significant technical advances. Start-up companies emphasize the applications of new technology and the introduction into commercial use. Technically oriented companies that are small cannot pursue either R&D or commercialization activities that call for substantial resources and complex organization. Large corporations probably should not become too involved in activities that do not.

The pressure to increase technical resources in pursuit of major technical advances may indeed result in more R&D people for each program in a large corporation and in fewer patents per person. Thus, a large corporation may have fewer numbers of marketable products per scientist or engineer than a small company. Clearly, that can be related to the complexity of the product.

A small company, particularly a start-up technology-based company, focuses on the marketability of a technical concept, not on R&D. Decisions are implemented more quickly. Starting up production can be rapid. Innovation in the sense of new applications on a modest scale can be highly effective in a start-up company.

This is why entrepreneurial activity has been an important factor in the rate of economic growth. Many areas of application are pursued in a relatively short time compared to similar actions of large corporations.

We need the Apples of our technology-based society to identify and pursue the economic benefits of technical advance. But Apple Computer needed the IBMs and Intels for the significant advances in computer sciences and microprocessors that made possible the economic introduction of a home computer.

An excellent chapter on this subject is contained in the book by Rothwell and Zegveld.[2] The critical point for this book is that the technical enterprise is made more effective by a system that encourages the creation of new technology-based companies *and* the interactions between small and large corporations.

FINANCING TECHNICAL CHANGE

Money flows from different sources to different parts of the process of technical change. The generation of science and technology is paid for principally by (1) corporate funds coming from revenues of the corporation to support the research organization of the corporation and (2) government funds—federal and local—coming from

taxes to support R&D relevant to needs of the public sector and conducted in government, university, and corporate laboratories.

A rough continuity exists on a macroscopic scale between these
funding sources and the magnitude of R&D expenditures. Each
industry sector is characterized by some percentage of sales devoted
to R&D, and government expenditures for R&D represent a fairly
continuous percentage of GNP.

On a microscopic scale, things are different. Individual companies
can raise or lower R&D budgets by substantial amounts in a given
year. Moreover, the direction of R&D effort can be changed significantly even when levels of R&D are relatively constant.

The amount and direction of R&D are influenced strongly when a
business plan for commercial exploitation can be set forth. Within
the large corporation, the integration of R&D, financing, and business planning is normal procedure.

The situation of interest here is when this transition from research
to use takes place outside the large corporation. External financing,
whether public or private, exerts an influence on the technical enterprise by performing several important functions.

- Independent business judgments aid in evaluating market objectives for new technologies.

- The need to minimize cash flow encourages rapid and close interaction among R&D, manufacture, and sales.

- Multiple sources of financing provide for simultaneous approaches
  to different applications of a technical advance.

These functions speed up the process of technical change and improve the overall productivity of the R&D process. A particular technical advance can have many possible applications in many industries. Each requires some additional technical effort, and each will
eventually require substantial investment for successful business
growth. The availability of third-party financing to identify markets
and to explore many parallel commercial applications (1) establishes
targets for technical organizations with related capabilities and (2)
identifies the opportunity for major investments needed in the successful approaches.

The new business activity supported by third-party financing
establishes linkages between R&D and the economic system. This
improves the probability of financial returns from new R&D pro-

grams. That is, it improves the selectivity process, and it improves the transfer of results from R&D to the marketplace.

Venture Capital

The successful use of venture capital in the United States to exploit technical advances through start-up companies is well established by the growth of the semiconductor industry, the recent explosion in applications of biotechnology, and in computers, peripheral equipment, and software. One of the great stories is of Silicon Valley, the area just south of San Francisco.[3] It demonstrates the involvement of venture capitalists with the network of new companies and new technical activities at universities, particularly Stanford.

Similar characteristics of the growth of new technology-based firms and the activities of venture capitalists were shown in the Charpie Report, named for Robert Charpie, chairman of the study group appointed by the Secretary of Commerce that examined development along Boston's Route 128.[4] The study was one of the earliest to ask: What made the Route 128 development so successful, and why had other regions in the United States with major universities and large financial and industrial communities not done as well? No simple answers are adequate, but the study revealed that a positive attitude evolved within the Boston financial community concerning the needs of new high-tech firms and the relationships that had to exist between the entrepreneurs and the venture capitalists. One circumstance fostering the bankers' support was their awareness that the New England region had lost an important economic contributor in the textile industry, so that the financial community was highly receptive to potential for new growth areas.

The U.S. experience in venture capital and its role in exploiting and diffusing technical advances is analyzed most thoughtfully by a banker from Great Britain, Matthew Bullock.[5] Bullock has since played a constructive role in the development of new high-tech firms in the vicinity of Cambridge, England.

The potential of high return on investment is the *only* criterion for venture capital. Assessing the attractiveness of start-up companies compared to other forms of investment is a highly subjective evaluation, however. The attraction is that a percentage of ownership, worth nothing at initiation, can be worth millions of dollars if the firm grows and prospers. The investors are individuals whose personal tax rates on ordinary income would normally be 50 percent (up to 70 percent a few years ago). Money earned from traditional forms of savings would be taxed at that rate, so that the saver would be left with just 50 percent of any earnings. On the other hand, profits gained by purchase of some asset, such as equity in a company, and then selling it after some legal minimum length of time are taxed at a much lower rate. Prior to 1970, this capital gains rate was 25 percent. Thus, an investor who sold out after the minimum period, formerly one year, would keep 75 percent of the earnings.

The effect of these relative rates was demonstrated dramatically in the period from 1969 to 1980. After the capital gains tax rate was increased in 1969, venture capital investments dropped sharply; the capital gains tax rate was lowered after 1977, and venture capital investments increased sharply. The numbers are given in Table 12–1.

Table 12-1.   Effect of Capital Gains Tax on Venture Capital Availability.

Period	Average Annual Amount of Venture Capital Committed ($ millions)
1969	$171
Capital gains tax rate increase	
1970–1977	58
Capital gains tax rate lowered	
1978–1980	596

Source: R. Rothwell and W. Zegveld, *Reindustrialization and Technology* (New York: Longman, 1985), p. 188.

Venture capital is intended to initiate a business, not to support R&D. Typically, a technical entrepreneur has tested the feasibility of applying a technical advance to a product or process, usually has demonstrated a working model or pilot plant, and very often has actually carried through initial manufacture and sales *before* a serious commitment of venture capital is made. An entrepreneur with a record of past successes may obtain venture capital on the basis of a business plan and technical concept, but this is hardly typical.

The potential returns from successful growth and lower capital gains tax rate may be sufficient to initiate new technical businesses once feasibility is demonstrated. They are not sufficient to attract venture capital to R&D. A business plan based on the *anticipated* results from R&D is simply too risky to justify investments under the conditions applicable to a start-up company. The fact that a lost investment could ultimately be deducted against future income, while a profit would be taxed at lower capital gains rates, still would not make investment in R&D attractive. Too many other alternatives for investment exist.

Hence, a new form of an older mechanism has come into increasing use since 1980 which can attract investment to support R&D. This is the limited partnership, well established as a tax shelter mechanism in fields such as oil drilling, real estate, and cattle raising. The common element is the policy judgment of Congress that it is in the public interest to attract investments to these areas. Since the risks are high compared to alternative investments, the investor would be able to deduct the investment from income immediately, thus "sheltering" it from taxes. Many changes are underway at this moment as to how much of the investment can be deducted in the year it was committed, but there is a substantial tax deduction at the beginning of the investment.

The public interest lies in encouraging a faster rate of commercialization from technical advance and hence increasing R&D activities that have the potential for commercialization. Pressure in these directions has been intensified in recent years by increased international competition plus the general need for economic improvement.

One obstacle to achieving these objectives has been that traditional venture capital is not available to support R&D which has commercial potential. A start-up company which has established technical feasibility is risky enough. However, an R&D limited partnership which permits immediate or early tax deduction can cut the investment cost in half. That is, an investor whose tax rate is 50 percent can save $0.50 in taxes for every dollar of investment which is deductible. Essentially, an R&D limited partnership becomes a form of tax-advantaged venture capital. The lower initial cost to the investor plus the treatment of future financial returns as capital gains, at a lower tax rate, *can* compensate for the higher risk involved. A business plan that is dependent on R&D adds a substantial technical uncertainty to the normal uncertainty of starting a new business.

The critical aspect of immediate tax deduction for investment in R&D rests upon Section 174 of the 1954 Federal Tax Code. This permitted a business to treat R&D as a cost of sales, so that the expenditures on R&D during a given year were deducted from revenues in that year before calculating taxes. However, Section 162 of the

Code specifies that such immediate deduction is for an "ordinary and necessary" expense related to the process of carrying on a business. On the latter basis, private investors who had sponsored or contracted for R&D had been denied any deduction since there was no "ongoing business." Financial support for R&D does not, by itself, qualify for a tax deduction, except to the extent permitted for charitable or educational purposes.

A landmark case was decided in 1974 for an investor by the name of Snow (*Snow v. Commissioner*, 194 Supreme Court 1876). He was a limited partner in an investment company which funded the development of a new incinerator and claimed a tax deduction for his share of the development costs in the first year, before a marketable product emerged. The Court ruled that the overall program included the express *intent* of developing a product, process or service for the market plan, and that Section 174 permitting immediate deduction of expenses would apply. Hence, support of R&D which was part of a broader business plan based upon the program could be treated as if the business already existed.

The first major R&D limited partnership of some public renown after the Snow decision was the 1978 Delorean venture to produce a new automobile. The engineering development of the automobile itself and the production of fiberglass reinforced plastic were supported with an R&D limited partnership which raised $18.75 million. In 1980, a Lear fanjet engine development was funded by a $30 million R&D limited partnership. Trilogy Computer Development Partners Limited raised $55 million as an R&D limited partnership for developing a large-scale high-performance computer system.

The early 1980s was a period in which the federal government, primarily the Department of Commerce, encouraged increases in R&D, particularly industrial research, as a basis for improving U.S. international competitiveness. A major emphasis was placed on expanding the use of R&D limited partnerships. Investor interest in the performance of "high-tech" industries was rising, buoyed by the sharp increase in commercial aspects of biotechnology. Finally, a growing familiarity with R&D limited partnerships spread among investment houses, sophisticated investors, law firms, accounting firms, and among the sources of technology—inventors, corporations, and universities.

As U.S. estimates of R&D limited partnership activity rose to a possible $2 billion annually, the potential impact of such partnerships on the effectiveness and direction of R&D nationally also rose. The Center for Science and Technology Policy at New York University initiated a program to judge the extent and nature of this impact, and determine the probable areas of constructive use for R&D limited partnerships. The report of this study, conducted principally by Lois Peters, is the source of data presented here.[6]

The number of separate deals to raise money for R&D limited partnerships reached a peak in 1983, but the amount of money raised continued to increase in 1984. The number of larger deals has been increasing. The data indicate that the funds raised for R&D limited partnerships are holding at roughly $1 billion annually. This is about 1 percent of all expenditures on R&D in the United States, and about 2 percent of industry-funded R&D.

It is instructive to look at the sources of money, since this touches on the public policy issues. The early large R&D limited partnerships were in the classical partnership pattern. A relatively small number of wealthy individuals and organizations put up money for a specific venture. The three mentioned above—DeLorean, Lear, and Trilogy—are in this group. The "limited" partners were financially liable only for the money put into the partnership, unlike the unlimited liability of a true partnership. However, they delegated management authority of the venture to the "general partner."

As the interest in the R&D limited partnership mechanism spread, specific ventures on a more modest scale were assembled by private investors—bankers, stockbrokers, lawyers, and accountants. Two-thirds of the deals were less than $20 million, with the majority ranging from $1 million to $10 million. Even a small deal for $2 million required participants of sufficient means so that they could invest, say, $100,000 each and afford to lose it all.

Large investment houses eventually established mutual funds for investing in R&D limited partnerships. These "blind pools" could now sell shares for $5,000 to individual investors, who would participate proportionally in the activity of the total fund. For example, Prudential Bache completed an $81 million fund in 1984. Merrill Lynch raised $70 million in 1985, and Paine Webber has offered a $100 million fund in October 1985, which had raised $47 million by January 1986.

Hence, substantial amounts of money are becoming available for R&D limited partnerships from investors of reasonable means. These investors may buy units worth $5,000 or $50,000 in a fund, but the risks will now be spread over 10, 20, or more separate partnerships. The fund allocates the resources to the many different ventures, and it acts as the general partner for each.

Since the funding of limited partnerships has many of the characteristics of funding start-up companies, it is of interest to note the distribution among technical areas.

Discussions with managers of the blind pools indicate that they anticipate investing in biotechnology, computers, hardware and software, medical technology, electronics, and energy products.

How significant a mechanism can the R&D limited partnership be in allocation of R&D resources, and should it be encouraged by public policy? A pragmatic answer is that if the level of funding of such partnerships stays near its present level, about 1 percent of industry-funded R&D, the results are all beneficial. The money appears to represent added money for R&D, though it may well have come in part from other tax shelter investments. It speeds up development of the commercial possibilities for newer technical advances, without removing money from more mature areas. Adding $500 million to test the feasibility of technical concepts should improve the effectiveness of venture capital which would follow in order to exploit the results through a continuing business. The extra technical effort would be too modest to have an impact on either the costs of R&D or the availability of technical personnel.

The R&D limited partnership can be applied to the several principal sources of technology—the university, the inventor, the private research institute, and the small technically based company with more technical-marketing concepts than money. And it even has certain advantages for the large company under financial strain or in a field of such rapid technical change that there is always more R&D to do when funds become available. But since it is intended to be an investment vehicle, not a device to support R&D, it may not be as appropriate for some applications as for others. It must protect the interests of the private investors, especially those in blind pools, who are often of moderate means. It attracts funds for R&D from a new source, but only by making the interests of those individuals and organizations the first priority. This is both a legal and pragmatic necessity.

One immediate requirement, then, is that the R&D limited partnership should provide some indication of technical feasibility in a short time, on the order of two to three years. This is not very patient money. Traditional venture capital supplied by sophisticated investors can be more supportive of longer term developments, even in a start-up company, than the R&D limited partnership. The second requirement is to establish a practical and credible business plan to produce a financial return following the successful completion of the R&D being funded. A third requirement is that the commercial potential of the technical project be high.

These conditions are not at all compatible with the university's needs or strengths. The inventor and private research institutes easily accept the need for quick indication of technical feasibility, but are not the best sources of business and market experience, and lack the capability for future exploitation of results. The large corporation has the technology, the experience and capacity for exploitation, but may bring in outside "partners" only under conditions that are not always optimum for the investors.

That leaves the small or medium-size technically based company as the most logical source and user of this means of financing. Such companies are sensitive to market needs, already have some technical, manufacturing, and marketing capacity, and possess the management know-how to add capacity. Most important for the investor, outside funds may be necessary to support very important technical programs. The financing may be beyond the current resources of the company, and delay can mean missing the opportunity completely. Other sources of funds would require either dilution of ownership in the core business of the company, or a debt that would seriously restrict profits.

Hence, the R&D limited partnership should continue to serve as a mechanism for aiding in the rapid growth phase of companies engaged in fast-changing technology, a phase when the funds can promote economic growth out of proportion to their actual amount.

It would seem that public policy should encourage R&D limited partnerships. Currently, uncertainty is reflected in debate on tax law changes that could lower the attractiveness of the mechanism for investors. As the gap between tax rates for capital gains and peak ordinary income is reduced, the added return may not appear to compensate for the added risk. If the deduction of the investment is spread out or the allowable amount reduced, the tax advantages begin to disappear.

These changes apply to other forms of investments as well. Even if the laws are changed, the attractiveness of R&D limited partnerships relative to other investments can be increased by making relations with the source of technology flexible, by permitting conversion of a partnership to equity, or by offering an option to participate in other development programs, for example.

The large corporation can make use of the R&D limited partnership for diversification to attract investors. The partnership may be a useful mechanism for collective industry research that precedes a business venture. Genentech has initiated limited partnerships on different occasions to pursue developments arising from its internal research.

Although most university research is too long term and costly to be an attractive risk for the typical investor, it is conceivable that large corporations might participate collectively in R&D limited partnerships dedicated to an area of university research relevant to their long-term business interests. The more probable university use of R&D limited partnerships may arise from certain institutions that have placed greater emphasis on the conversion of knowledge to economic use. Carnegie-Mellon University, Rensselaer Polytechnic Institute, and Georgia Institute of Technology, among others, have focused research programs on the scientific and technological needs of industry. The partnership mechanism might well prove to be effective in providing funds for the final, expensive stages of development, resulting in long-term royalties to the university as a limited partner.

Whether the university should ever be a general partner is debatable. The commercial obligation this entails is in direct conflict with other university objectives. Even as the R&D source and contractor, two categories of desirability are present from the university view. If investors have bought shares in a blind pool, which then negotiates a limited partnership with a university, there is at least a buffer between the investor and the university. If an arrangement is made with the university by an investment group *before* all the funds are raised, there is a strong possibility that individual investors will participate *based* on the university's reputation. Disappointment in the event of commercial failure could develop resentment among influential citizens against the university, a most undesirable state of affairs for a nonprofit public or private academic institution.

Venture capital financing for start-ups or through partnerships in tax-advantaged programs can influence the technical enterprise but represents just a small proportion of nationwide R&D expenditures.

It does, however, provide a psychological drive and focus that is at least as important as the money. It fine-tunes the American technical system not just in market development and new technical applications, but in morale, in excitement, in public interest. And it does speed economic growth. These facts should be taken into account when public policy calls for reexamination of tax laws. The private interest of independent investors has served as an important tool in the implementation of major national objectives through the growth of the technical enterprise.

TECHNICALLY BASED REGIONAL ECONOMIC DEVELOPMENT

Very important, creative, and effective actions are occurring today in regions—states, cities, metropolitan areas—which are targeted to geographic boundaries, specific companies, and resources, and are significant in the direct relationship developed between the technical enterprise and regional economic development. Although regional economic development is not a new form of economic behavior, planning development around technical advances to the extent currently pursued is a vivid example of many aspects of the technical enterprise discussed in this book.

Regional development is where much of the action is today. It is the concrete application of what, to most political leaders and the general public, must be considered as abstractions. The nature of "scientific and technological progress" or the "process of technical change" may form the intellectual basis for a book, but those concepts do not stir political or economic action. The immediate objectives of regional authorities are (1) jobs for the residents of the region and (2) tax revenues from industry and commerce operating within the region.

Regional authorities everywhere appear to be convinced that "high-tech" industries represent growth opportunities. They initiate many programs designed to attract such activities and expand those already located in the region. Planners in regions dependent on mature process industries, which have lost markets and jobs in the 1970s and 1980s, note the example of Route 128, in Massachusetts, with technology-based companies more than replacing the jobs and tax revenues of textiles. They see the growth of California's Silicon Valley and the success of the Research Triangle in North Carolina.

How can these examples of growth, prosperity, and stability be followed? Can technical advances be used for regional economic development, and what must be done to do so? Quick answers are that

- Economic benefits can be derived from regional activities to encourage technically based industry, but expectations must be realistic.

- Much can be learned from past examples in order to modify and apply, not follow.

- Specific actions must be based on an understanding of how the system works.

To discuss these issues in more detail, it is necessary to separate actions aimed at technically based economic development from general economic development. Every region tries to improve its attractiveness for business and for people. This translates into concern for taxes, services, security, transportation, education, and cultural institutions. All the items that make life more pleasant and business more profitable are stressed. I want to look more carefully at features which are of particular relevance to technically based industry, assuming that all other economic and personal amenities are being pursued.

One important element is a "critical mass"—the grouping of facilities and people with interests in technical change. Accumulating this mass takes time, usually a much longer time than is estimated by regional planners. Consider an example.

The origin of Silicon Valley goes back to the electronic developments of Lee de Forest and others at the Federal Telegraph Company in Palo Alto from 1913 to 1930. The modern era started perhaps with Hewlett-Packard in the late 1930s, followed by Varian Associates and Ampex. A major electronics complex was already in being by the mid-1950s, with 53 member firms from that region in the Western Electronics Manufacturers' Association in 1955. The emphasis on semiconductors grew after Shockley Transistor Corporation began in 1956, and several of that group left in 1957 to start Fairchild Semiconductor. At least 30 years of growth in semiconductors can be counted, based upon a preceding 40-year base of high-technology electronics companies.[7]

Much of the activity in technically based regional economic development is associated with local universities. The more and better the educational institutions in a region, the greater the attraction for residents. Benefits result for different groups, and any regional program must be tailored for the particular constituency in mind. There are several categories of benefits.

- Convenient and lower cost education for undergraduates who live in the region is a general benefit for all economic groups but especially those of limited means.

- Opportunity to obtain a graduate degree by part-time study is of great value for junior professionals in all industries and has special appeal for firms with substantial numbers of engineers.

- Adult education courses offering cultural enrichment and practical background subjects enhance the quality of life for the citizens of a region.

- Specialized training and graduate degrees can be tailored to the region's industries.

- Faculty competence in technical areas relevant to regional industry can provide both consulting and the opportunity for joint research efforts.

- Specialized technical facilities and libraries can support technically based firms.

- Major research facilities and centers of competence offer valuable linkages for technically based firms in related fields.

This wide range of benefits is not exhaustive. The intellectual climate they represent is an attraction to most people and hence to most companies. However, these benefits are either indirect and institutional in character or, when specific to a need, are long-range. The role of universities in technically based regional economic development is very rarely one that has an impact in a few years.

Two exceptions from this statement are of great current interest. One is the economic benefits attached directly to a major research facility. The other is the growing regional concern with entrepreneurial activity.

The era of big science after World War II taught political leaders and regional planners a valuable lesson. A major research facility, whatever its indirect benefits for future economic growth, is an important economic benefit in its own right. Just as the economic attraction of defense business itself is high without regard to economic carryover into the civilian sector, so the conduct of R&D can be an economic good for the region without any regard for the ultimate value of the output.

For over 30 years in the United States, regions have competed for major research facilities launched by the principal mission agencies, and, indeed, by private corporations. Whether the installation is a NASA Space Center, a defense research laboratory, a National Laboratory of the Department of Energy, or a research laboratory of IBM or Monsanto, the local impact is good economically. With costs of roughly $100,000 to $150,000 per professional person, a research facility with 100 scientists and engineers means a payroll of about $12 million annually; a large facility with 1,000 professional personnel brings in about $120 million annually. Add on hotels, restaurants, and other local services that follow, and the regional benefits are obvious.

Not surprisingly, it is accepted that political pressures are exerted from all regions whenever a large government research facility is at issue. In a healthy democracy, either the political jockeying takes place on behalf of technically qualified regions or so many regions are vying for selection that the politics is neutralized and technical qualifications become the basis for selection. A third mechanism of site selection is old-fashioned logrolling, trading votes so that one politician's pet project is approved in return for that of a colleague being approved at a later date. This oft-maligned practice has resulted in a geographic distribution of facilities that is probably less concentrated than would have resulted from choice on technical merit alone, and this is certainly not an evil result.

Thus, traditional political practices used for highways, dams, post offices, and military bases did not cause too much outcry when applied to large government R&D facilities. One area, however, has until now been off-limits for such political pressures. This is the proposal for a specific research activity at a university. Research proposals are normally evaluated on an absolute basis—that is, should the research be supported on its merits?—not on the basis of which university should receive a grant.

In 1983 and 1984, however, a series of proposals from universities for research facilities received the funds through action of Congress, not from the agencies which would normally sponsor research to be conducted in these facilities. Admittedly, approving facilities is in a different category than approving actual research programs. Nevertheless, a plan for facilities is based upon a statement of the need for a particular area of research, an action plan to develop programs and staff, and possibly a statement of research intent. Presumably, it makes little sense to approve the facilities unless one approves the reasoning behind them. Thus, congressional action in a noncompetitive situation does indeed bestow implied congressional blessing on a research plan, something which the legislators themselves would surely agree they are not qualified to do.

That is not, of course, the way the issue would be stated by those requesting congressional help or by Congress. There should be a clear distinction between the desire for regional development via research facilities and the requirements for the best research and the generation of significant technical advances. This is a hot issue currently in the technical community. The issue developed in 1983 when Columbia University and Catholic University were awarded $5 million each in congressional legislation for new chemistry laboratories and a new laboratory for vitreous state studies, respectively. Apparently, both universities had retained political advisers. There was also congressional action for $20.4 million to the Oregon Health Sciences Laboratory for an information center, $15 million to the University of New Hampshire for a space and marine sciences building, and one or two other grants. During 1984 Congress approved $7 million toward a supercomputer center at Florida State University (planned to be a $63 million facility paid for 70 percent by the federal government); $19 million for an engineering center at Boston University; and more. This is reported in an article by Colin Norman in *Science* under the unkind, but probably accurate, title of "Pork Barrel Scoreboard."[7]

Philosophically, the issue is captured by the contrast between the buzzwords "peer review" and "comprehensive merit evaluation." *Peer review* refers to the practice of NSF, NIH, and other organizations which support largely university research to solicit evaluation of proposed research from other experts in the field, based solely on the technical merits of the research and researchers. *Comprehensive merit evaluation* is the term applied to the process used in obtaining political support, described at a meeting of the National Academy of Sciences in July 1985 as recognition that "major research facilities . . . inevitably affect the economic, social, environmental and other circumstances of the regions, states and communities in which they are located."[8]

The case for comprehensive merit evaluation was put forward in a paper at the NAS meeting by John Silber, president of Boston University. He points out first that the debate is over "15 Congressional actions [totaling] approximately $100 million to help 15 universities build new or improved facilities." Also that "facilities have for the most part never been subject to the traditional peer review," and that peer review is used in only 8 percent of federal R&D activity. (Since defense R&D was about 60 percent of federal R&D in 1984, peer review would therefore apply to about 20 percent of nondefense federal R&D.) Silber observed that "20 institutions . . . in three geographic regions . . . receive nearly half of all federal research support," and that this perpetuates an old-boy network. In an exchange with William Carey, executive officer of the American Association for the Advancement of Science, Silber referred to the "mythical notion of peer review as some form of immaculate conception whereby with precision and great objectivity, decisions of merit are made." He asserted that the value of the term *comprehensive merit evaluation* is that "it acknowledges complexity where complexity exists." All of this has been described by Daniel Greenberg, who summed up the controversy by saying that "the threadbare institutions have no loyalty to a system that has consistently excluded them as unworthy."[10]

One can hardly call Boston University or Florida State "threadbare." However, it seems reasonable to assume that a peer review system tends to make the strong research groups stronger. With deliberate efforts and establishment of guidelines, the peer review system can encourage and develop lesser groups. Though imperfect, it has worked well by the simple test of the diversity and vitality in the U.S. research system.

What is clearly disturbing to the AAAS and the NAS, is the prospect of extensive replacement of this system by political judgments, stirring memories of Lysenko in the USSR. That would indeed destroy U.S. science. It is incumbent upon the leaders of science and technology policy in the United States to keep reminding us all of the critical need for objectivity and high standards in the support of technical programs.

Beyond this, John Silber's case is reasonable and should be considered in more depth. Encouragement should be given to institutions where technical growth can stimulate broader interest in education and research and where that growth provides a nucleus for regional economic development. Approving a small percentage of expenditures for R&D, focused primarily on facilities, on regional considerations would not endanger the broad base of technical competence. This requires members of Congress to be more familiar with the directions of U.S. research, and this improved knowledge can have considerably broader value for the future support of technical activity.

The opportunities and the complications in combining congressionally approved technical plans with regional economic development are well illustrated by the program of Northwestern University in Evanston, Illinois. The city, working with the university to develop a new research park that would attract new industry, has planned a $400 million development in the park. To get a nucleus and first tenant, the interested parties obtained support from their congressmen, Sidney Yates, who also chaired an appropriations subcommittee of the House. That subcommittee prepared a bill, approved by the House, to provide $26 million for a Basic Industry Research Laboratory at the research park. It is intended to "increase energy efficiency in manufacturing and the conservation of energy" by basic industry such as manufacturing and metal fabrication.[11] The University is establishing a Basic Industry Research Institute which is expected to attract tenants to the park. In perhaps 10 or 15 years, an estimated 8,600 new jobs will be created, partly with the help of an "incubator" facility to stimulate new businesses.[12]

These are substantial numbers, with the $26 million of federal funds a key element. The complications lie in the development of research programs. The legislation anticipates that funding will come from the energy conservation programs of the Department of Energy, which had not planned for it. If support for operating funds is mandated by Congress, as seems likely, then other research could be affected.

The involvement of Congress in technically based regional economic development is a risky path to be walked with great caution, careful guidelines, and constant vigilance. A ceiling on funds and avoidance of judgments on research programs would provide some limitations to such action. When used judiciously, it can encourage sound development activities. If not, we could experience a new breed of white elephants that would very likely bring the practice to a halt in due time after wasting the efforts of technical personnel and great quantities of taxpayers' money. In the short run, of course, the region benefits directly from construction and from jobs in the new federally funded facilities.

Regional development in recent years with respect to high-technology industry has placed much emphasis on entrepreneurial activity. One common approach is to offer financial incentives and often physical assistance to provide low-cost space for small companies. Renovating old textile mills and factories in New England is representative of this activity.[13] Providing modest space and, very often, some technical services for budding entrepreneurs still in the development stage is another constructive approach, often referred to as "incubator space." A popular location for this type of facility is at or near a university. Presumably this encourages faculty to explore commercial possibilities of their technical advances, and simplifies technical cooperation with university staff and facilities for any occupant of the space.

In general, the role of universities in entrepreneurial activity is very real yet mixed. It is clear that one strong indication of successful economic output from university research and faculty is when such activity is initiated by the university itself. This has been a common characteristic in the successful entrepreneurial activity related to Georgia Institute of Technology, Rensselaer Polytechnic Institute, and Carnegie-Mellon University. The University of Pittsburgh has initiated a Foundation for Applied Science and Technology to facilitate the flow of relevant technology to the "marketplace through partnership with industry" and, according to its chancellor, Wesley Posvar, is entering "into arrangements like those with more hopeful expectancy than apprehension."[14]

It is equally clear that these relationships do not occur automatically. A region cannot, or at least should not, expect that the simple presence of a university ensures the blossoming of new high-technology companies. Entrepreneurial activity is influenced considerably by the philosophy of the university, its administration, and the peer pressure of faculty. The atmosphere of interest in technology-based industrial growth which emanates from a Stanford University or Massachusetts Institute of Technology stimulates more entrepreneurial activity than any specific program. This atmosphere is not present in many other institutions, often by choice. The diversity of American universities is strikingly evidenced in this one area. A region does not "impose" entrepreneurship on a university.

Financial support for new firms, both venture capital and loans, is the quintessential element of capitalism, yet here also, regional activity is playing a role. This is often done through establishment of a public non-profit corporation to provide such funds or drawing upon a state employees pension fund. Four states were doing this in 1979, and roughly 25 in 1985. The emphasis has been on high technology firms, a source of irritation to other types of small businesses in the states.[15]

A more direct form of regional financing has been initiated by the town of North Greenbush, New York, near Troy, the home of Rensselaer Polytechnic Institute. RPI has a 1,200-acre high-technology park in the town. Drawing from a $750,000 grant for economic development from the Department of Housing and Urban Development (HUD), a venture fund of $100,000 was set up. The town has provided some funding for eight start-up companies in which it now owns differing amounts of equity.[16]

Such regional activity in financing is no threat to Wall Street, but does raise questions of criteria. Private venture capital, however tough-minded, focuses on market potential and financial return. Publicly supported venture capital adds some further considerations, notably jobs in that region. These are not incompatible, but there will inevitably be trade-offs, and hence reason to set up independent review and guidance. Surely some ventures will be approved because of job potential, but with greater market uncertainty. More firms will be given a chance, which is good, but the overall productivity of the technical enterprise may stop. It is a matter of probability, and the extent of regional venture capitalism is unlikely to be a cause for serious concern.

One final comment may be appropriate on technically based regional economic development. Emphasis is placed on the economic opportunities associated with technical advances. Since new products, processes, and services should result, there is a probability of economic activity and employment. The amount of employment is occasionally a basis of misunderstanding.

At first glance, high technology means new jobs. At second glance, much sober analysis indicates that not as many jobs will be associ-

ated in new biotechnology and semiconductor companies as anticipated, certainly not as many as in the growth of services generally
or that may be lost in mature process industries. This can lead to dis-
enchantment with the prospects for new growth.

But yet another glance should be taken and should be followed by
careful analysis in order to tailor policies for regional development.
Three levels of economic activity can be associated with technical
advances, and they broaden considerably with significant technical
advances:

1. Activities in the forefront of technical advance, by the companies
    engaged in engineering and development of microprocessors,
    computer peripheral devices, laser-based communications—the
    homes for new Ph.D.s in biotechnology and computer science.

2. Activities in existing industries which absorb and apply new tech-
    nical advances. Here we have the major process industries which
    use results from biotechnology—food, chemicals, pharmaceuti-
    cals; the service industries which are revolutionized by micro-
    chips and computers—banks, publications, telecommunications;
    the automobile and metal fabrication industries affected by
    robotics and computer-aided engineering.

3. Technical services and support industries. This category includes
    instrumentation and specialized technical analysis; unique pack-
    aging and transportation requirements; specific components,
    maintenance, construction.

The second and third levels of impact from new technologies can
have very considerable opportunities for employment. New York
City, for example, as the heart of financial and telecommunications
activity, is perhaps the largest user of the microelectronics revolution without benefit of producing the chips. Every region has its own
opportunities for drawing upon high technology.

Regional programs that facilitate the absorption of new technologies and development of specialized support services are perfectly
valid tools for economic growth. The Center for Science and Technology Policy, at New York University, conducted a study supported
by several regional authorities on possible actions to stimulate biotechnology commercialization in the metropolitan New York area.[17]
We pointed out the growing needs of local major process industries
(chemicals, food, and others) for people trained in biology, chemical

engineering, and computers in order to modify production processes and the opportunities for local universities in developing their curriculum. The instrumentation and analytical services required for these uses would assist the major industries of New Jersey and the instrument and electronic firms of New York's Westchester County and Long Island.

Thus, regional economic development is at the forefront of the technical enterprise. It represents the practical border between technical advance and economic use. Because it involves the average citizen and the responsible political leaders, it is probably the most effective activity for integrating the technical enterprise into our society.

ENTREPRENEURIAL EXPERIENCES OUTSIDE THE UNITED STATES

This chapter would be incomplete without some mention of activities in Europe and Japan. Oversimplifying, there is an increase in American-style start-up businesses and venture capital activity in both Europe and Japan today, but the nature and role of new technical firms in these regions will remain much different than activities in the United States.

In late 1982 I discussed this subject in Tokyo with a friend who was then the senior technical officer and member of the Board of Directors of one of Japan's largest electronics corporations. I expressed my curiosity over the fact that, whereas in the United States the creation of new technical firms has been an important factor in the initiation of technical concepts and their introduction into the marketplace, Japan, while highly successful in such technical innovation and marketing, did not have much activity on the part of new firms. How did they provide the functions carried out by new firms?

His answer was that the Japanese practice of lifetime employment in large corporations (at least until age 55) provides a strong incentive to initiate new product concepts. With some number of people in excess of what would be needed for current operations, the company encouraged the identification of potentially valuable developments and would assign excess people to them. Essentially, these exploratory programs were conducted with zero charges for the costs of technical personnel. Thus, new developments can be conducted under the umbrella of a large corporation at little or no cost for a long period of time.

In the past, it has proven more difficult for a small new company in Japan to become a supplier to large established industries. This is true as well, of course, in the United States. A purchasing agent who buys a personal computer from IBM will not be held at fault if the item does not work properly, whereas if he buys from an unknown firm and the product is defective, his judgment may be questioned more seriously. The added obstacle in Japan is the web of relationships among groups of large companies, wherein long-established supplier–user linkages characterize a "family" of companies. This, too, may be changing as new products and technologies emerge, but it has been an inhibition to entrepreneurship.

To a lesser degree, Europe is influenced by the security and status of large company employment. While such security attraction is equally important in the United States, there does seem to be more psychological importance attached to status in Europe. The title of "Chief Engineer" at Thomson in France, or "Group Leader for Organic Chemistry" at Bayer in Germany bestows identity, respect, and honor. The title-holder's family, neighbors, and former classmates associate a social and economic status with the title and with the company. The stability and importance of the company transfer to the individual.

In the United States there is perhaps more respect for the American who "starts a company" and who, therefore, becomes his or her "own boss" rather than an employee or hired manager. Failure at such an endeavor is not judged harshly by friends and colleagues. In Europe, however, the act of breaking away from a large company still raises questions of loyalty as well as judgment, and failure may not be treated as kindly.

Efforts to change all this, to encourage entrepreneurship, and to set examples of success are underway throughout Europe. The emergence of a cluster of new companies near Cambridge University has been noted, and England seems well on its way to overcoming any psychological hang-ups related to new business formation. Government agencies, major banks, and private venture capital are all active.

While the Continent has been somewhat slower, much is stirring there also. Venture capitalists are more active in France, Germany, and Sweden than they were as recently as 1980, and with a number of entrepreneurs as clients, particularly in businesses related to computers and software. A major difficulty has been the lack of a large market equivalent to the American over-the-counter (OTC) stock-market. This makes it difficult for the venture capitalist to take a new company public, sell stock to a larger group of investors, and thereby provide a return to the venture capitalists.[18]

Hence, despite courageous and commendable efforts, Europe has not yet succeeded in achieving the easy vitality of start-ups in the United States or the encouragement of exploratory developments by the large Japanese corporations. The countries of Europe, with the encouragement of both socialist and conservative governments, are attempting to remove obstacles to the use of venture capital. Government-funded venture capital agencies are being used by many European countries, quite similar to the financial mechanisms used by state governments in the United States.

The question in all such activity is the effectiveness of mechanisms that expedite the conversion of technical advances to economic use, diffuse technical change throughout industries, and guide the emphasis of technical efforts on the basis of probing markets and developing specifications.

We in the United States have learned to make use of venture capital and start-up companies. This fits our social structures, our economic system, our organization of R&D activity. The Japanese appear to have mechanisms for performing these functions within the structure of the large corporation, a difficult feat for most large U.S. corporations. Europe is struggling to produce comparable mechanisms for the encouragement and economic use of new technical concepts. The results will very likely not be quite that of the United States or Japan. It is probable that more government involvement will be present in funding and in equity holdings. Some combinations of government agency and large corporation sponsorship may provide an umbrella for shepherding new ventures. Whatever mechanisms are proven to be acceptable and effective, they will be European. The technical enterprise does not require uniformity for growth and efficiency but it does need linkages between the generation and use of technology. These appear to exist or are developing.

NOTES TO CHAPTER 12

1. Bernard Rothlein and I received our Ph.D.'s in physics together at the University of Pennsylvania.

2. R. Rothwell and W. Zegveld, *Reindustrialization and Technology* (New York: Longman, 1985), ch. 6.

3. "California's Great Breeding Ground for Industry," *Fortune*, June 1974, p. 128.

4. U.S. Department of Commerce, "Technological Innovation: Its Environment and Management," 1967.

5. M. Bullock, *Academic Enterprise, Industrial Innovation and the Development of High Technology Financing in the United States* (London: Brand Brothers and Company, 1983).

6. L. Peters and H. I. Fusfeld, "R&D Limited Partnerships and Their Significance for Innovation," Center for Science and Technology Policy, Graduate School of Business Administration, New York University, 1986.

7. Ibid.

8. *Science 226* (November 2, 1984): 519.

9. *Science and Government Report 15*, no. 13 (August 1, 1985): 2.

10. *Science and Government Report 15*, no. 13 (August 1, 1985): 2.

11. *Science 225* (September 28, 1984): 1454.

12. *Chemical and Engineering News*, November 5, 1984, p. 6.

13. *New York Times*, "Signs of Economic Revival Abound in Northeast," June 23, 1985, p. 1.

14. Wesley W. Posvar, "New Horizons for the University," Editorial in *Science 225*, no. 4669 , September 28, 1984.

15. *Wall Street Journal*, "Development Aid from States Is a Growing Factor for Firms," October 8, 1984, p. 33.

16. *Wall Street Journal*, August 7, 1985.

17. L. Peters and H. I. Fusfeld, "Biotechnology in Metropolitan New York and New Jersey: Resources and Future Growth," Center for Science and Technology Policy, Graduate School of Business Administration, New York University, December 1983.

18. *Fortune*, "Europe Rediscovers the Entrepreneur," October 2, 1983.

13 TECHNOLOGY, SOCIETY, AND PUBLIC POLICY

The technical enterprise exists to serve society. Because it is a cause of change, society adopts a variety of public policies to set conditions upon the technical enterprise. Controls on the export of technology are one example. Regulations to reduce risks to health and safety are another. Any deliberate action of government to use or to restrain some particular technology certainly has an impact on the operation of the technical enterprise in that area. But other parts of the technical enterprise are affected and policies may have unintended effects.

Science and technology are not pursued in isolation. They are connected closely to all other activities of society. The numbers of new students entering technical careers, the availability of funds to support technical activity, interest in particular technical subjects all reflect the needs and the conditions of society.

Suppose the growth of the technical enterprise rested only on generating knowledge, developing applications which produce new techniques and new demands, then generating more knowledge related to these developments. We would then have an expanding spiral of interactions that would soak up all the people and resources in society. Long before that happened, of course, we would run out of people who wanted to engage in R&D, and resources would be restricted by other demands of society. In fact, however, more practical mecha-

273

nisms operate within society that slow down the process of technical change.

One such mechanism is the time lag required for a significant technical advance to be absorbed and utilized. The knowledge inherent in any new advance is available almost immediately throughout the technical enterprise. This can have an effect on research programs and even development concepts in a fairly short time. Nevertheless, the diffusion of a major technical advance to the point where it is embedded in many products and industries requires working familiarity with both principles and practices by those responsible for those products.

Consider the spread of the transistor. Following 15 years or more of increasing basic research on semiconductor materials, a demonstration was held at Bell Laboratories in 1948 of a working transistor. It was described as a device "which has several applications in radio where a vacuum tube ordinarily is employed."[1] Presumably a period of several years was required for professors of electronic engineering to become thoroughly familiar with circuit designs incorporating the transistor, and for textbooks to be prepared. By the early 1950s, beginning students of electronics could go through their full course of education with a thorough grounding in transistor circuitry. These students emerged with a B.S. or M.S. by 1960. In another four to six years after graduation, a substantial number of this new generation were in positions where they were responsible for product development and design, possibly as chief engineers or product line managers. It was not until then, the mid-1960s, that the surge of electronic products and systems based on transistor circuits began to move into the marketplace. From 1955 to 1970 total sales for semiconductor devices increased from $40 million to $1.3 billion annually.[2] Thus nearly 20 years passed from the innovation to broad commercialization.

A similar time lag seems to be occurring in the adoption of personal computers. From about 1978, college students have gone through their degree studies in constant companionship with computers. These students entered the work force in professional capacities in the early 1980s. As this cohort become parents and managers, we could expect a marked rise in sales of personal computers for homes. The major personal computer marketing effort thus far has achieved only mediocre results by trying to persuade middle-class, middle-aged Americans to use computers for routine tasks. The industry should expect far greater sales to the economically rising 30-year-olds, who would feel lost without a computer easily available.

Similar developments are occurring as well in biotechnology. Many decades of basic research provided the knowledge and ability to alter genetic material and thereby produce useful drugs and chemicals by biological processes. This opens the door to the development of materials with properties that cannot easily be accomplished by other techniques. New applications in treating disease, in agriculture, in foods, and in other areas are being explored intensively. Importantly, manufacturing processes in these industries which now produce these substances by chemical processes can, in many instances, be replaced by biological processes. This calls for technical personnel trained in biological techniques, particularly fermentation, and concerned with the organization and control of process manufacture.

This overall expansion and diffusion in biotechnology has spread through the technical community. It has been reflected only modestly in its impact on the marketplace, in the actual sale and use of products based upon these new techniques, in actual large-scale manufacturing processes within such major industries as food, drugs, and chemicals. The potential exists, and all the background steps are in progress. However, the complete technical and manufacturing base requires years for development, for training, for large-scale manufacturing.

For better or worse, society is providing this time in part through the necessary lag in the educational process and, particularly in biotechnology, through regulatory processes.

Regulatory processes provide a form of friction in our economic system, adding time and cost to the process of technical change. This is not at all a deprecatory statement. Friction in mechanical systems or resistance in electrical circuits serve important functions in our ability to control these systems. We pay a price by their consumption of some part of the energy available, but they are invaluable as well as inevitable properties in the controlled operation of a physical system when properly designed. Zero friction is not only impossible physically but would permit systems to run out of control too easily. But too much friction uses up too much energy, reducing the efficiency of the system.

In the development of biotechnology, regulations serve to slow down technical innovation in a planned manner. General oversight of techniques used in handling recombinant DNA material has been in the hands of the Recombinant DNA Advisory Committee of the National Institutes of Health. Guidelines were devised in 1976, and new research experiments are reviewed in advance by the Committee to see that they conform to those guidelines. The EPA has taken the position that they have the same responsibility for products and processes based on genetic engineering as they do for more traditional chemical products and processes, and intends to devise such regulations. The Food and Drug Administration still has the primary responsibility for those products and considers products from the new techniques in their purview. Clearly, coordination is needed and regulations are an important factor in expanding this new area.[3]

The objectives of regulatory action by society are to reduce risk and improve the quality of life. If we define "risk" broadly enough, that definition probably includes regulations on behavior. Our concern here is simply with actions that affect the technical enterprise.

Presumably, society intends to reduce risk at some commensurate cost. Two difficulties arise immediately. One is that there is a trend in law and in rhetoric to try to *eliminate* risk, to reduce it to zero. The other is that the concept of "commensurate cost" is ignored, since cost is extremely difficult to measure and is highly subjective. The combination of eliminating risk legally and ignoring cost consequences can be catastrophic.

Every technical advance can be the basis for producing benefits and for producing some form of damage. I do not want to become diverted to ethical questions, but for the purposes of this discussion I must take the position that knowledge and understanding are neutral. The fact that basic concepts in physics made it possible to produce an atomic bomb, or that genetic engineering raises the uneasy thought of influencing characteristics of future generations must not inhibit research for greater understanding of basic physics and genetics.

Regulations, then, concern the use of knowledge, although the voluntary restraints and guidelines for research procedures in biotechnology have been a form of exception to this rule. These did not prevent the search for knowledge, though they did restrict the methods.

The practical issues are somewhat less global than avoiding atomic bombs or specifying the population of the twenty-fifth century. They deal with emission standards for chemical plants and municipal waste disposal, with automobile fuel consumption, with the safety of a punch press and of children's toys, with health effects of food additives and side effects of new drugs.

These concerns for health and safety are necessary obligations of government—that is, of society. Their application introduces added costs and time into the process of technical change. It is not clear that the impact on this process is properly taken into account by legislators and regulatory bodies, particularly with regard to other programs that could have been pursued by the technical community. Fewer new drugs introduced, fewer new products generally, less effort available for increased productivity. All of these are costs to the consumer and to the economy.

The point is that regulatory actions, however necessary, slow the process of technical change. This is a serious cost and should not be imposed lightly. By slowing economic growth, adding to consumer costs, and reducing the competitive position of U.S. industry, just as friction absorbs energy in physical systems, regulations absorb economic activity. Society has accepted the obligation of reducing risk in the introduction of technical change, to avoid unacceptable harm even as we encourage pursuit of benefits. It can be a stabilizing factor in setting an orderly time period for acceptance, during which the technical base can be properly prepared.

There is no such thing as zero risk. Society and its political representatives must accept and treat this fact realistically. The true costs of unrealistic expectations are unacceptable, if we take into account the improvements and benefits that would be held back by an impossible goal.

A distinction must be drawn between harm resulting from lack of knowledge or even unavoidable conditions and harm resulting from willful disregard of known hazards. Lawsuits and resulting damages related to mishaps from products and processes are having adverse effects on the technical enterprise. The willingness to introduce a new material, product, or process is being restrained beyond prudence or responsibility.

The normal safety net of liability insurance is becoming less available and too costly due largely to the precedents being established in U.S. courts. Insurers in Europe who normally participate with U.S. insurers to provide liability coverage for U.S. industry are now withholding such participation because of the U.S. system of tort law. This is an extremely serious limitation on corporate operations that involve major technical change.

Liability insurance is one mechanism by which society can absorb acceptable risk of harm while encouraging technical advances that can create benefits. When this mechanism is endangered by the courts, it is a legitimate topic for consideration by legislators who must balance all national interests. This is the case in many states regarding medical liability, where doctors are withdrawing from certain fields of practice because the premiums have increased so sharply. Maximum limits of liability may be set by law.

Any such applications to the process of technical change must be considered very carefully, so that responsible actions will still be encouraged while irresponsible behavior is prevented. Still, serious restrictions on our technical capacity will produce risks of their own to the nation's economic and political health. That is surely a consideration for society in trying to minimize risk to our individual welfare.

PUBLIC POLICY AND THE TECHNICAL ENTERPRISE

Certain public policies are intended to direct or limit the operations of the technical enterprise. These policies may take the form of fairly specific regulations, or they may be reflected in more general actions. An example of the specific kind is the current emphasis on export controls. The more general form of public policy is the debate over industrial policy. Both are important components in the interactions between society and the technical enterprise.

Export Controls

Export controls on technology from the United States are intended to prevent access on the part of a potential adversary nation to any U.S. technology that can be used for military applications. Both the definition and implementation of such controls pose enormous difficulties, and they have a very significant impact on the technical enterprise worldwide.

There is little quarrel with the *intent* of export controls. Common sense indicates that no great social or intellectual purpose is served by shipping a device for undersea detection of submarines to the

USSR, or a system for laser-guidance of missiles. Those are the easy decisions. The issues quickly become more complex when the question is one of selling landing systems for Soviet airports which embody advanced memory chips, available in Europe, or exchanging technical data between American and Soviet metallurgists or, as in recent instances, refusing permission for technical personnel from NATO countries to attend a professional meeting on advanced optical systems.

Any restriction on linkages within the technical enterprise weakens the effectiveness of the overall system. Thus, a deliberately restrictive policy should be based upon careful analysis of the local gain or loss within the technical enterprise from the restrictions. This requires a sound grasp of how the system works. Several complex issues must be addressed.

First, the objective of export controls should not be viewed as keeping important technology from potential adversaries. Instead, the objective must be to keep ahead in those technologies. In some areas, this leadership may be maintained by restricting exports. In others, it may be improved by free trade and technical exchanges. Restriction of exports is only one tool in a broader objective.

Restricting the flow of important technology to a potential adversary requires a control over the worldwide technical enterprise that simply does not exist. The vast number of linkages and of independent technical activity in all countries create a global sea of science and technology, not just a few easily controlled channels. Focusing on those technical developments in which the United States has the leading position at a given instant, restrictions on the export of such technology *anywhere* would have to be considered in order to restrict it from the Soviet bloc.

It is impossible to identify an area of science or technology as having only commercial or only military applications. Every technology or scientific advance has what is known as "dual-use" capability.

Restricting a particular technology does not affect only that device or that area of know-how. It can affect our future capacity for generating new advances in that or related areas.

Here is precisely the core of the difficulty. Accept the broad objective of export controls—to achieve and maintain a technical advantage in critical technical areas over a potential adversary. If any given decision affected only a particular device or specific know-how, we could resolve all gray areas on the conservative side, and restrict its export. However, when we limit the operation of the technical enterprise, we limit our own progress. The judgment is therefore more than just the normal complexity of whether the device or know-how can be of military aid to an adversary. It involves the far more difficult task of deciding whether, by preventing some short-term help to a potential adversary, we are limiting our own long-term capabilities.

There is no question in time of war, since the short term is the only term that counts. There is a very serious question when we must think about a very long, indefinite period during which our economic strength, based upon advancing technology, as well as our military capabilities, for which we want a broadly superior technical base, can both rely upon our ability to continue significant technical advances in science and technology.

The breadth of controls on technology has two aspects. The controls are not simply on finished products or processes but on technology. While the intent of export controls is to limit the export of particular technologies to a potential adversary, generally considered to be the Soviet bloc countries, regulations necessarily apply to exports of those technologies to any country.

Given the objectives of an export control system based on national security needs, it is thoroughly logical that the system should consider technology generally and the flow from the United States to any external source. The very logic of these requirements dramatizes the fundamental conflict that they represent with the effective operation of a technical enterprise that depends increasingly on a network of linkages, alliances, and strong interactions with international markets.

The system by which the United States attempts to control exports was established during World War II and has been in use since then.[4] Much of the present concern comes from the reauthorization of the Export Administration Act of 1979. The implementation of the Act, administered by the Department of Commerce, focused traditionally on the shipment of something physical—a product, a device, a report, a blueprint—something that embodied a particular technology or that described it in specific terms. A license must be issued by the Commerce Department for such shipments.

The definition of technical data used in the Export Administration Regulations (EAR) refers to "information of any kind that can be used or adapted to use in the design, production, manufacture . . .

of articles or materials." Technical data is not limited to physical devices but can be know-how conveyed in discussion, in a lecture, or during a visit. Indeed, after 1979, individuals and companies were reminded that a license is necessary to export technology in any form outside the country and that this could include discussion with foreign nationals visiting in this country or by U.S. citizens traveling overseas.

In order to maintain some sense of proportion, so that licenses need not be obtained to discuss material in texts and public documents, four lists are used to guide those who may need licenses as well as those who have to provide the licenses. The first two are a Commodity Control List and a U.S. Munitions List. These set down a large number of materials and items which must be considered for licensing by the Commerce Department. A third list is used by the Coordinating Committee on Export Controls (COCOM) of items that should not be exported to Soviet bloc countries. While COCOM is broad (Japan plus the NATO countries minus Iceland), the interpretation and enforcement of the COCOM list is not uniform.

These three lists are public information. A fourth list in preparation for some time by the Department of Defense is classified, but a descriptive table of contents has been published. It is a 700-page listing called the Militarily Critical Technologies List (MCTL), and sets down those fields and subfields of science and technology that can be relevant in differing degree to military applications. Since the Commerce Department must consult with the Defense Department on questions of national security, presumably this list will be an important reference for such consultations.

The enormity of the enforcement problem is obvious. Most companies today are often in unintended violation of the literal application of export regulations with regard to communication and visits with colleagues from friendly countries. Even though attempts may be made to limit the refusal of a license to the most advanced, most critical technologies, the process for application must be undertaken by a very great number of companies for a great number of situations. Apart from the time and cost, which can be excused if the objectives were achieved, there are other serious negative consequences for the overall system.

The first, and most critical, is the potential slowing down and weakening of the U.S. technical base. The sources of technical change in each company are varied. While the largest activity is very likely that of the domestic R&D organization, a substantial amount of input can come from foreign subsidiaries, joint ventures with foreign companies, licensing from overseas, technical exchange agreements, research interactions with universities. These linkages can all be affected by export controls. Further, the magnitude of R&D conducted by a company depends upon the sales of its products. If substantial markets are cut off by export controls, this would not only result in less money flowing back to support R&D in those areas, it would stimulate and strengthen competitive R&D overseas. Those competitors would not necessarily take as strict a position on export controls as the United States for the same type of products.

This leads directly to the second consequence, namely, the impact on our relations with friendly countries.[5] Putting aside political issues, reactions occur that affect technical developments and trade. Senior personnel in England from several of the major British companies assert that one of the top priorities of their R&D programs is to achieve independence from the United States with respect to advanced components and systems in electronics. They have expressed deep concern that items from the United States on which they depend for telecommunications and data processing might be cut off by U.S. export controls. Similar comments have been made by French executives and those involved in the ESPRIT program of the European Community.

Such reactions can shift the relative balance of technical strengths internationally on specific components and systems. They will almost certainly affect the overseas markets of U.S. companies. The results may, in the long run, be exactly the opposite of what is intended by the system of export controls.

A system of export controls of technology contains the implicit assumption that one has control over that area. The worldwide technical enterprise is too strong, diverse, and flexible for that assumption. It works automatically to overcome the effect of obstacles and delays to technical advances.

Does this mean that export controls should be abandoned? No, but they must be realistic. Controls can be more effective if their scope is limited to a specific device or product. Even then, the effect is to slow, not to prevent. Any broadening of controls reduces their effectiveness. Most important, there is a cost to very action, and that cost must be evaluated realistically with full knowledge of how the technical enterprise operates. Any broadening of export controls beyond a specific device with obvious military implications drops off sharply in effectiveness and increases sharply in cost to domestic capabilities and leadership. It should become almost self-limiting in its coverage, provided a solid basis for evaluation is developed.

Two studies to do just that were initiated in 1985. One is a substantial broad inquiry by the National Academies of Science and Engineering that will consider trade effects and priorities for alternative control measures.[6] It focuses on industry, following an earlier study by the Academies of university research[7] known as the Corson report (for the chairman of the study) Dale Corson, president emeritus of Cornell University). A second study is by the New York University Center for Science and Technology Policy. It is narrower, and focuses on "Impact of Export Controls on Industrial Research."

Again, the real objective of government concern must be to develop and maintain technical advantage over a potential adversary, not simply to keep helpful technology from being exported. This does not mean that we should be careless or foolish about placing useful military devices on the open market. It does mean that the cost and manpower required to broaden controls might be used more effectively in strengthening our technical base. Certainly when we consider the scope implied by attempting to restrict knowledge of new advances, including possible restrictions on foreign graduate students and limiting attendance at professional scientific conferences, we decrease our capacity to generate those advances.

Staying ahead is also the more pragmatic approach. I know the actions needed to make technical progress in a field and to translate the progress into useful products and processes, both civilian and military. The United States has developed a strong and productive system for generating and converting technical change. I really do not know how to prevent those results from being known and used by others except in the most obvious and limited instances. But I am certain that if a potential adversary has to obtain those advances from us, then we are more advanced and will very likely stay that way. Any change in a system which produces that result must be subject to the most serious and thoughtful analysis.

Industrial Policy

An example of public policy which affects the technical enterprise more generally is that loose collection of actions and guidelines which can be referred to as "industrial policy," the position of a government with respect to the mix of industries desired to meet the national objectives. If a critical national objective were to be self-sufficient in energy or materials, that leads to a particular industrial policy. A desire to emphasize labor-intensive products, or high-value-added products, or to emphasize exports can each be the basis for industrial policy. The objectives of full employment and favorable trade balance do not identify an industrial policy, since they are common to all situations.

The industrial policy of a country may be manifested in different ways, it may not be internally consistent, and it may change over time. The official U.S. position is for minimum government involvement in industry and for free trade, but specific actions restricting imports of steel and automobiles have the effect of an industrial policy, though unstated. Japan has shifted from labor-intensive industries to high-technology, high-value-added industries. France has demonstrated the desire to have a strong and independent base in information technology. Sweden and the Netherlands have focused on exporting.

Whatever the industrial policy of a country, many tools are used by government for its implementation, but R&D is increasingly the focus. A government's choice of an industrial policy immediately defines the objectives for that country's science and technology policy with respect to economic growth. The attempt to formulate science and technology policy in the absence of a consensus on industrial policy leads to inconsistencies and ineffectiveness, as many responsible for U.S. science and technology policy in the legislative and executive branches can testify.

Some industrial policies are stated explicitly though not necessarily consistently. There seem to be more logical relationships between industrial policy and government actions in science and technology on the part of other countries than there is in the United States. This may be because the combination of large markets, natural resources, and a strong technical base gives the United States the luxury of not having to select an industrial policy that optimizes limited resources.

Thus luxury may not last, since pressures are building in the United States that are not too different from those faced by our industrial trading partners of the OECD. One principal issue being forced upon us for various reasons is the appropriate balance between the mature process industries, which have been declining in employment, and the high-technology areas, which have shown rapid growth rates of employment prior to a semiconductor slowdown in 1984-85. This issue blends into three related issues which seem to be saying the same thing differently, but actually do bring in additional factors:

- Low-technology versus high-technology industries
- "Sunset" versus "sunrise" industries
- Service industries versus manufacturing industries

However this issue is described, the technical enterprise plays a role on either side. The entrepreneurial activity in high-tech companies is popular in all countries. Even though mature process industries have difficulties, no government is attempting to hold back high-technology growth, despite possible short-term unemployment effects. The universal approach is to encourage whatever growth is possible from domestic high-technology efforts; otherwise the results will simply be imported.

The decline in total output and employment in process industries is worldwide. More capacity exists than the world market requires. Increased productivity due to new processes and controls also means less employment for any given level of output. Hence, much of the painful impact in some industrial sectors is basically a readjustment to international markets.

Nevertheless, much activity is stirring in different forms that is stimulating the technical enterprise to strengthen process industries. This ferment is evidenced in three categories:

1. Management of manufacturing is the focus of much attention in business schools and engineering schools. Specialists in management techniques coupled with hands-on experience in the plant are entering industry and can serve to shepherd in those advanced technologies relevant to improve productivity and change approaches to manufacturing processes.[8]

2. Introduction into the process industries of technical advances emerging from the rapidly changing areas of microprocessors, computers, robotics, biotechnology, and others.

3. Allocation of increased technical resources to the manufacturing processes of these mature industries.

Much recent investment in the United States for expanded and modernized plant has been based on the classical combination of new technology and capital investment, the latter stimulated by tax incentives such as an increased investment tax credit and accelerated depreciation.

New technology is being drawn upon in computer-controlled steel-making, in reinforced rubber produced in automated plants, in automated looms fed by robotic materials-handling carts within the textile industry. The technology does not solve all the problems of these industries, but it improves the competitiveness of those corporations making the investment.[9]

Regional economic development emphasizes the technical needs and opportunities related to regional industry. The University of Michigan is attempting to work with the heavy manufacturing industries in the state through a combination of appropriate training and cooperative research. Its president, Harold Shapiro, advocates the strategy of focusing on the skills and capacity attached to the high-wage workers in the state, rather than shifting to low-wage sectors.[10] Northwestern University has a program to initiate a major research institute for technical support of basic industries characteristic of the midwest.

The attention to technology in traditional manufacturing is international. A major concern of the United States and Europe has been the early and increasing emphasis on technically advanced production processes in Japan, which both lowers costs and improves quality. The approaches being taken among the "survivors" within British industry have begun to strengthen companies there such as Babcock International, Tube Investments, GKN, British Steel, and British Leyland. Both product innovation and improved manufacturing productivity are playing a role. The experience there, as in the United States, shows that unemployment is not immediately helped, but the increased competitiveness in world markets should result in employment gains in the long run.[11]

It is very possible that the deep concern everywhere concerning employment may, in fact, shift more attention to the low-technology and the mature industries. This is not because their growth opportunities are suddenly expanding but because some disillusion is setting in about the employment growth in high-technology industries. The current decline in the semiconductor industry has shaken some of those concerned with regional development who placed their confidence in it. The United States has witnessed this on the West Coast in particular, so that the need for diversified industry, even the scorned "low-tech" industries, is a practical requirement.

This has been reflected in U.S. plants overseas, where cutbacks in semiconductors have been exacerbated by the lack of adequate support industry to take up the slack, and without satisfactory efforts in transferring technology to local industry or training local workers. The need for more industrial options is a growing basis of government policy.[12]

If there is any guideline in the changing mix of industrial policies, it is that a strong and diverse technical enterprise can support whatever needs and opportunities arise. Moreover, the networks that characterize the technical enterprise provide precisely the flexibility necessary to respond to the changing economic and political factors which set the emphasis for industrial policy at any period.

This has been evidenced in the shifting emphasis between encouragement of high-technology industry and support for process industries. Each country or region attempts to do both when this is possible and relevant. The available technical institutions have an obligation to work with political planners so that realistic programs can be initiated. Such activities have become a major feature of the U.S. university system today.

One particularly interesting example of the interactions between industrial policy and the technical enterprise emerges from observing the cluster of "high-technology" countries composed of Sweden, the Netherlands, Switzerland, and Israel. Each has a limited domestic market, yet each has succeeded in developing world-class technical competence in selected areas. This requires a concentrated technical base to provide high-value-added products, and broad international markets to support the technical effort. Each country is unique, but there are common threads of encouragement and direction of the technical enterprise to achieve the results.

COMMENTS ON INTERACTIONS WITH SOCIETY

This chapter has been essentially a reminder that the technical enterprise, as part of a broader society, is influenced by actions of that society with respect to direction of effort, level of funds and personnel, and procedures. Thus, the built-in processes of society have as much to do with the rate of technical change as the technical activities themselves. The effectiveness and productivity of R&D are influenced not only by the selection processes and linkages established within the technical enterprise but by the obstacles or encouragements of public policy.

A continuous feedback, however, weaves a much more complex relationship between society and the technical enterprise. What society does is dependent on what it is capable of doing. The divestiture of AT&T would not have been contemplated without the technical advances in telecommunications that made competitive systems and investments practical. Export controls seem more important to national security because of the significance that advanced technology has for electronic warfare, yet the breadth of the technical enterprise that led to these advances defeats the effectiveness of these controls. Our concern with protection of copyrights on creative materials in the media, written and electronic, arises from the rapid technical change in the ability to copy and communicate these materials.

In more philosophical terms, the level of technology defines the level of civilization. Hence, the actions of society to regulate its own behavior inevitably translate into pressures upon its technical base.

Cause and effect are not easily identified. The technical advances of the next decade will determine the nature of our society and the directions of economic growth well into the next century. Equally, the operations of our economy and the characteristics of society provide the organization, the support, and the needs for the technical enterprise which encourage the present allocation of technical resources and pattern of technical activity.

NOTES TO CHAPTER 13

1. *The New York Times*, July 11, 1948.
2. Richard C. Levin, "The Semiconductor Industry," in *Government and Technical Progress*, edited by Richard R. Nelson (New York: Pergamon Press, 1982).
3. "Government Readies Rules for Biotechnology Control," *Chemical and Engineering News*, August 13, 1984, p. 34.
4. Mitchell Wallerstien, "Scientific Communication and National Security in 1984," *Science 224* (May 4, 1984): 460.
5. The observations in this paragraph are based on comments made to me in discussions with senior executives in the United Kingdom and colleagues

in France and elsewhere in Europe who are involved in the EC's ESPRIT program.

6. "Academics Study High Tech Export Controls," *Physics Today*, August 1985, p. 48.

7. "Scientific Communications and National Security," NAS/NAE, 1982.

8. R. Hayes and W. Abernathy, "Managing Our Way to Economic Decline," *Harvard Business Review*, August 1980.

9. "Can Smokestack America Rise Again?" *Fortune*, February 6, 1984.

10. Harold Shapiro, "Michigan Can Manufacture A Solid Future," *Detroit Free Press*, September 25, 1984.

11. Roger Eglin, "Britain Can Make It," *The Director*, August 1984.

12. "Enthusiasm for High-Tech Industries Is Waning in Malaysia," *International Tribune*, July 10, 1985.

14 HOW TO USE AND STRENGTHEN THE TECHNICAL ENTERPRISE

The twentieth century has witnessed the emergence of the technical enterprise as a major factor in society, and a powerful tool in economic and political affairs. It is much more than a convenient name for the sum of all the groups of scientists and engineers who conduct research and development. It is a structure which, through concentration of resources and its networks and cooperative mechanisms, makes possible the continuing advance of science and technology within the finite resources of the world economy. By drawing upon it constructively, both industrial and governmental objectives can be furthered.

The wonders of biology and electronics, of jet planes and communications are often praised, but corporate executives and political leaders cannot simply order more of such outputs. They can only influence organizations, relationships, and funding. Voters and taxpayers cannot dictate technical progress, but they can decide on the conditions for carrying out these activities and the allocation of public funds.

Statements about technical advances and society do not connect those advances to the decision-making processes of industry and government. The operations of the technical enterprise constitute that connection. We can and do influence the mechanism by which technical change is generated, not its actual generation. To do so realistically and effectively requires an understanding of which mechanisms are most likely to produce the desired technical change and how. That is much of what this book has emphasized.

The industrialized countries have come through a fascinating and unique period since World War II. Nearly half a century has been marked by new relationships between technology and society. Acceleration in the growth of science and technology was encouraged by two factors. One was the maturing of industrial research, which linked organized R&D to substantial funds and integrated it with marketing and manufacturing. The second was the ability to conduct large and complex technical programs, using the systems management developed for military applications.

Industrialized society is familiar with the process of technical change so essential to the business plans of corporations. The expectation of technical change is built into the objectives of government, particularly defense, but also in communications, energy, environment, transportation, education, and others. Familiarity raised expectations and led to dependence.

During these years there was in the United States a sense of technical self-sufficiency. *But this period is ending.* We now rely upon continued technical advances without the confidence that each organization, public or private, can provide the technical resources necessary to produce those advances.

Decisions in the real world are not based on philosophy but on pragmatic approaches to immediate problems. The increased cost and complexity of continued technical advances have led to interdependencies and linkages.

This is a major characteristic of the period we are now entering, the third phase in the evolution of the technical enterprise.

In phase 1, *User Independent of Generator*, prior to the late nineteenth century, the technical enterprise was largely outside of the principal user organizations. Universities and individual research and invention were the sources of technical advances. Industry, commerce, and government drew upon and modified relevant technical change for their separate objectives.

In phase 2, *Combined User-Generator Growth*, from the late nineteenth century to the present, the technical enterprise grew increasingly *within* the principal user organizations. The combination of user and generator of technology accelerated the rate of technical advances and provided relative technical self-sufficiency for industrial growth and national objectives.

We are entering phase 3, *Relationships among User-Generators*, when growth of the technical enterprise will depend on linkages among technical units of different users and with all sources of technical advances.

The inexorable pressure on resources needed for continued technical advances, and the resulting structural changes in the technical enterprise, pose great challenges to our technology-based society. The reactions to thousands of situations that are affected by those pressures are creating answers to these challenges through ad hoc decisions. The pressures themselves would rarely be articulated, perhaps not even recognized as a fundamental condition affecting the worldwide technical enterprise. A principal objective of this book has been to bring about more effective reactions to these pressures by the widespread recognition of how the technical enterprise works and how it is changing.

CHALLENGE

One immediate challenge is to industry. Continuing technical advance calls eventually for resources beyond the capacity of a single company. This is producing many forms of linkage—joint ventures, R&D limited partnerships, consortiums, collective research associations. These traditionally noncompetitive mechanisms have not weakened competition to date, and there need *not* be any decrease in competitive behavior or in the continued generation of technical change.

The technical enterprise represents a powerful tool through its potential for concentrating technical resources in a massive technical information network and mission-oriented programs.

Each corporation can draw upon external resources at some price, whether direct fee, staff time, or restrictions on the use of technical advances that emerge from any group support. The wider range of technical choices strengthens the corporations that participate and provides a stronger technical base for the country or region that supports a collective R&D program. Besides resources devoted to R&D, commercial success depends on many other factors—market size, selectivity, feedback, conversion to use, internal linkages between R&D and other corporate functions. Nevertheless, it is reasonable to assume that the corporations that participate in the national or regional collective R&D programs are stronger because of that participation *provided* they maintain their independence for competitive action. It is that independence which increases the overall productivity of R&D with respect to economic growth, not simply the ability to produce some technical advance.

This is the heart of the industrial research system, which couples technical progress to the introduction of technical change. The technical enterprise makes possible, through intelligent use, a concentration of technical resources that *extends* the capability of any individual corporation. The corporate structure and competitive drive provide the discipline of feedback and selectivity that increases R&D productivity experienced by society as a whole.

Consider the duplication or redundancy represented by the fact that a number of corporations may be pursuing a similar technical objective, whether it be a stronger steel, a more powerful microcomputer, or a genetic approach to food preservation. Each company adopts a somewhat different research program. Duplication lies in the similar, perhaps identical, market objective. The total effort may encompass many technical programs, all differing to some degree. But the research phase of the process of technical change is relatively low cost. As each program proceeds, some will be dropped because of technical difficulties. Those which prove technically feasible will enter the more costly phase of engineering or product or process design. Some of these will be cancelled due to the difficulty or cost necessary to produce a practical design. The remaining programs which appear capable of reduction to practice now must be subject to prototype manufacture for a product or pilot plant test for a process. These are the truly expensive phases of R&D.

The competitive industrial research system encourages duplicate technical approaches in the early stages of high technical uncertainty and relatively low project costs. As the program proceeds to the higher cost development and engineering, technical uncertainty decreases. Once a program enters a $10 million prototype or a $100 million pilot plant stage, there should be almost no uncertainty left about scientific or engineering principles. Little, if any, duplication remains at these stages.

That is the basic discipline of competitive industrial research in a market economy. Mistakes in principle or in costly approaches are filtered out in the low-cost stages. Major investments, while facing market uncertainties, can proceed with confidence with regard to the technical base. Any effort to select a "best" approach too early in this process can carry a costly technical concept through to the final product or process. This results in an inherently uneconomic or uncompetitive result. That is indeed the danger of a planned economy, or occasionally of an overzealous industrial policy within the OECD countries.

The combination of competitive industrial research with selected collective R&D through the technical enterprise can be a powerful mechanism for continued technical advances. If we weaken the competition, we reduce R&D productivity. If we ignore the potential value of the technical enterprise, we raise costs or reduce technical advances or both.

Society benefits from the internal operations of the technical enterprise and from its interactions with those external activities that represent the many objectives of society.

The experiences of the past 40 years have taught something about the relationships between the user and the generator of technical advances which are valid for public programs as well as private. A focused mission, a concentration of technical resources, and feedback between user and generator are all positive factors in the process of technical change. Thus, when society in the form of government is the user, as in defense or space, its relations with the technical enterprise are analogous with those for industry. The flow of science and technology, the use of relevant collective R&D, even the encouragement of multiple research approaches to determine the most effective product or process for major investments, are all mechanisms that can improve the productivity of R&D in the public sector when government itself is the user.

IMPLICATIONS

The new phase of the technical enterprise must inevitably influence actions of industry and government intended to stimulate and exploit the process of technical change. What can we expect and how can the interactions be made most productive?

Every corporation tends toward a balance among its resources and objectives. Given the increase in the technical resources needed for continued technical advances, the balance between R&D and corporate objectives will be maintained by actions in one or more of the following categories.

1. Market niches should be emphasized that provide opportunity for economic growth through engineering design, applications of

technical advances, and modest extensions of technological know-how. These can be pursued with current levels of R&D activity. The approach is traditional for small and medium-size firms, but may be characteristic increasingly for large firms.

2. Technology niches should be emphasized that provide opportunities for economic growth through the generation of significant technical advances in narrowly defined areas that can form the base for new materials, components, systems, or processes. These may be pursued with some increase in the present levels of R&D activity. This approach will be appropriate for large firms that possess substantial technical resources in areas of rapid change, so that a concentrated effort is possible in a few, but not all, sectors of the expanding technical field.

3. Corporate participation should be increased in collective R&D activities that strengthen the technical base of relevant business interests but that do not limit the corporation's independence with regard to new business development.

4. More joint ventures should be formed for the pursuit of new business opportunities as a mechanism for expanding resources.

The overall trend is for increased specialization in niches, requiring an increase in collective efforts in order to develop and exploit major business opportunities that require competence in several technical or marketing specialties. The larger the corporation, the larger the specialty may be. However it is defined, a growth opportunity outside that specialty will increasingly be pursued through linkages rather than organizational additions within the corporation.

The increased readiness of corporations to expand their technical base through external linkages offers considerable opportunity and obligations for university research. There is a substantial gap between the conduct of traditional "undirected basic research" at universities and the focused programs of industrial research guided by the feedback of commercial exploitation.

The major American research universities are playing an important role in filling this gap through their enlarged function as managers of research. This is evidenced principally through the organization of mission-oriented research centers that conduct "directed basic research." To be effective two things must happen:

• Closer working relationships should be formed with industry to obtain the benefit of feedback.

- Structural changes at universities must be pursued to make it possible for such research centers to appoint and retain research staff so that missions have continuity.

Research centers at major universities can serve as a bridge between industrial research and the broad base of largely undirected university research activity. As feedback and planning become more effective, the emergence of commercially valuable patents and know-how becomes more probable. This in turn can provide stimulation and support for research activities among faculty not tied in formally with the centers.

Thus, the healthy growth of mission-oriented research centers at major universities can serve on the one hand to strengthen the technical base of industry and government while encouraging the more traditional less structured university research activities. To do this well is a considerable challenge to universities, but the ultimate benefits for both society and for academic research are far greater. The fundamental requirement is that each university engaged with such research centers must recognize explicitly that it has accepted a new function as a manager of research, not simply taken on an appendage that adds more research funds, and that this function requires separate treatment not necessarily tied to teaching or departmental controls.

Governments of the industrialized nations have become increasingly attentive to the general problem of how to draw upon their technical resources in order to achieve certain national objectives. Not that many years ago, this attention was devoted largely to the increase in basic research activity at universities. Undoubtedly this added to the reservoir of scientific knowledge, increased the number of technical graduates, and strengthened university research. The impact on economic growth was not at all direct or rapid, and in most fields was nonexistent.

Today, more realistically, government actions are more concerned with supporting economic growth. Action falls into three categories:

1. Government support of industry initiatives in the form of direct research grants, matching research funds, and tax incentives.
2. Government encouragement of, and support for, university-industry research cooperation.
3. Government facilitation of, and support for, collective industry research programs.

Less than 20 years ago, government policies on technology and economic growth emphasized government initiatives in specific program areas, with little concern evidenced for industry planning and capabilities. It was common to refer to the "adversarial" relations between government and industry. These may still exist in regulatory areas and in political philosophy throughout the OECD countries. It is disappearing in the realm of science and technology policy due to the clear evidence of industry's role in strengthening the effectiveness of each nation's technical resources.

FUTURE CONCERNS

Two important strengths of the modern technical enterprise permit technical progress to continue, even accelerate, despite the need for steadily increasing resources to maintain a constant rate of significant technical advances. These strengths or mechanisms are:

1. A widespread pattern of technology flow through a vast number of networks and channels, public and private, so that the rate of diffusion and absorption of scientific and technical advances can partially offset the need for increasing technical resources.
2. The development of many forms of linkages, public and private, which provide the capability for concentrating technical resources in areas of common interest and on specific technical objectives.

These twin abilities—to diffuse technology rapidly and effectively and to concentrate technical resources through external linkages—are principal criteria for judging the impact of emerging forces within society. There will surely be exceptions, but a sharp warning signal should go up whenever either of these capabilities is diminished.

Consider the current development of a system of *export controls* to limit the availability of U.S. technical advances that might strengthen the military base of a potential adversary. The approaches and difficulties of export controls were discussed in the preceding chapter. The issue here is the relationship between important user-generators of R&D, the large technology-intensive U.S. corporations, and the networks within the technical enterprise. The network's effectiveness is diminished by not receiving the full inputs from the generators. This is precisely the intent of export controls and is presumably an acceptable price if the objective of national security is achieved. However, the largest generators are also the largest users of technical advances. Any flow of technical information through the technical networks has the greatest value to the largest users, since there is a higher probability for successful absorption and conversion. In brief, the United States provides the greatest inputs of new science and technology into the technical enterprise, but it is in a position to receive the greatest value from it. The irony of the export control system in its broad application is that the most technology-intensive country with the most technical advances to protect can be the most adversely affected by diminished technology flow. Only by the most limited, highly selective application of export controls can its highest priorities be achieved.

This is not solely an American phenomenon. The Alvey and ESPRIT programs both contain restrictions on the export of their technologies. Such actions could expand if the U.S. controls intensify.

Less direct but even broader adverse impacts on the technical enterprise would result from any substantial increase in *protectionism* within the world trading system. Limitations of imports by the industrialized nations would shift the focus of industry from widening international markets to the unnatural restriction of domestic markets. The immediate effect would be a smaller revenue base to support R&D thus reducing the capacity to concentrate major technical resources for continued technical advances. This would decrease the generation of science and technology in all countries, cutting back technology inputs throughout the technical enterprise.

Reduced markets and the consequent reduction in R&D could in principle be incentives for greater technical cooperation to maintain an adequate technical base. This, however, assumes that industrial needs and manufacturing capabilities are not receiving adequate technical inputs for exploitation. If market growth and strategic business planning result in sharply lower ability to use new technology, then the momentum for more technology weakens, and the need for external linkages is reduced or eliminated.

An area which has the potential for adverse impact on the technical enterprise, particularly in the United States, is the ongoing process of *"restructuring"* within American industry. This results from the so-called takeovers of major corporations with presumably undervalued assets, particularly in the area of natural resources, or with a mix of businesses whose separate values if spun off add up to more than the market value of the company's stock. The consequence of such takeovers, or of actions initiated in advance to avoid being vulnerable to a takeover, often involve a trimming away of product lines or a narrower focusing of business interests within an industry. This in turn is accompanied by a reduction in the amount of R&D expenditures and in the breadth of relevant technologies and exploratory technical activity undertaken.

None of these consequent actions are wrong per se, and in many circumstances they can be quite correct and overdue. Certainly the optimum use of economic resources as judged by the marketplace and the protection of shareholders' equity are two strong and proven foundations of the private enterprise system. And clearly, R&D that cannot be exploited effectively is a poor use of technical resources.

Yet there is a nagging uneasiness that such actions can reduce the momentum and the strength of U.S. industrial research if the phenomenon of restructuring goes too far. Possibly this uneasiness arises because the stimulus for the actions arises from defensive moves, from an inward-looking strategy, a kind of drawing the wagons into a circle, and not from an unhurried deliberate plan for growth. Industrial research has flourished in an expansive, outward-looking, aggressive posture. While this has produced some ineffective R&D efforts, new laboratories that were later closed, show case R&D that was poorly planned or integrated, it also produced in the United States the strongest technical base of any country and the greatest single force behind the modern technical enterprise.

This is good reason to look carefully at the current so-called restructuring, even from the narrow viewpoint of R&D activity alone. There is a high degree of national interest present in any serious shift of motivation and of financial support for industrial research.

This of course has not happened on any broad or systematic level. The U.S. technical base is not shaken. The technology-intensive industries—electronics, chemicals, instruments, drugs—have not been involved in unfriendly takeovers.

But the uneasiness and potential for damage are there and must be monitored. The oil industry has been the subject of much restructuring, driven by the impact of lower world oil prices. Hence, Phillips Petroleum is reducing its technical effort, a direct fall-out from the takeover activity of T. Boone Pickens and others. Exxon, following a relatively "clean" business decision to close out diversification activity through Exxon Enterprises and at Exxon Office Systems, is refocusing its technical activity on its core business and traditional technologies. Whether this results in cutting back its recent broad expansion into biotechnology as a new, but relevant, technical competence will be evident in 1986 or 1987. The purchase of Crown Zellerback by interests headed by Sir James Goldsmith has been followed by indications that it will close or drastically cut back its corporate research laboratory.

The gradual conversion of the American Can Company to emphasis on financial services, and the purchase of the Continental Corporation, have resulted in a lessening R&D activity. The not unfriendly mergers of Stauffer Chemical with Chesebrough Pond, and of Allied-Bendix with the Signal Corporation will surely produce a decreased overall R&D effort. The Uniroyal Corporation, taking the initiative to refocus and streamline, has indicated plans to trim back R&D accordingly. The attempt by GAF to take over Union Carbide, whether successful or not, will very probably coincide with a cutting back of that chemical giant's research efforts.

The difficulty inherent in all this is not that total R&D is reduced. The sum of all the reductions to date is probably on the order of $500 to $600 million, roughly 1 percent of all industry-funded R&D. It is not with the personal impact on individuals or communities, painful as they are.

Rather, our long-term concern is with the potential decline in our capability for concentrating technical resources. This is, after all, perhaps the single greatest U.S. competitive advantage in technically based industrial growth. It relates to the size of the firm's revenue base which, in the United States, derives from a large domestic market.

A major U.S. corporation may be involved in several industrial sectors. Even if it is almost entirely in a single sector, there is a spread over many product lines or segments within that broad sector. In either case, the corporate research capability is supported by the sum of all these revenue sources. A major effort may be pursued that will be exploited in a single product area, but the technical resources devoted to it draw upon the overall corporate resources.

It is this characteristic of modern industrial research which is endangered by any widespread restructuring. A realistic abandonment of declining businesses can strengthen the overall corporation. But a deliberate narrowing of corporate interests based only on the price/earnings ratio of stock can weaken the ability or the willingness of a corporation to conduct major technical programs in either a new or a mature business area. This would be a serious setback to a principal factor in U.S. international competitiveness, the breadth and momentum of the industrial research base.

THE FUTURE PRODUCTIVITY OF R&D

Private and public institutions which generate or convert technical advances will adapt to the characteristics of the technical enterprise by changing their practices and structure. A more pragmatic way of stating this is that each organization will carry out its objectives with the optimum use of the resources available. Given the steady pressure for increased technical effort to produce significant advances, the practices and structure of R&D organizations will change to accommodate that pressure, either through deliberate planning or practical reaction to specific problems.

These gradually shifting characteristics of producing technical change can be viewed as improving R&D productivity. Improvement is difficult to measure, but its significance is straightforward. It refers to obtaining more useful output from a given amount of resources devoted to R&D.[1]

Pragmatically we can only consider the R&D productivity of a particular laboratory or technical organization with respect to its unique situation. Thus, a productive research program at Monsanto is very unlikely to be considered productive for IBM, and a university research center is judged by wholly different criteria of productivity. R&D productivity is principally influenced by three factors *that are controllable*: selectivity, external interactions, and management of technical resources.

Selectivity. Picking the right project, the right approach, can be critical to making a timely contribution and meeting particular research objectives. In an industrial laboratory, selectivity means relating the research to the capabilities of the corporation and the technical base of the relevant industry. For a graduate student at a university, it means choosing a project in which meaningful data can be obtained within a one- to three-year period. In all cases, selectivity is based on access to the technology flow within the networks of the technical enterprise. What is desirable and feasible depends on what exists in the technical community and what is possible to generate with the resources of the technical enterprise.

External Interactions. Productivity of R&D at a particular laboratory is enhanced by the ability to extract data and concepts from other technical organizations, to cooperate when appropriate in ongoing R&D, and to provide for effective transfer of a technical advance to other organizations that can build upon it—design and engineering groups, manufacturing, other technical units within the host system, private or public. This transfer process is expedited by the growth of communication networks within the technical enterprise. It is increasingly valuable as the technical community becomes more familiar with the many working relationships that can be established. The technical enterprise must not be viewed simply as a beneficial, though passive, reservoir of science and technology, but as a mechanism that can be used in the process of technical change. An elaborate system of electronic message exchange is being used by the Alvey Programme in the United Kingdom to connect the 106 companies, universities, and research institutions that participate, typical of advanced networking.[2]

Management of Technical Resources. With selection of the "right" project and increased awareness of the technical values available from the technical enterprise, productivity of the R&D process then derives from "bread-and-butter" R&D management. This is the optimum use of available resources to implement the technical objectives. It involves such factors as the proper mix of professional researchers and technicians, the investment in research equipment, and the use of external resources.

The growth of the technical enterprise affects all of these factors through its broad communication networks, the rapid interplay between theoretical concepts and advanced experimental techniques and the growing acceptance of using *all* technical organizations worldwide as supplementary resources, when available, to the internal technical resources of the laboratory. The technical enterprise offers more options to the R&D manager in the resources available, hence for increasing R&D productivity.

The technical enterprise formed by the growth of R&D activities worldwide provides mechanisms for technology flow and technical linkages which permit that growth to continue with available resources. Thus, the technical enterprise emerges as an important factor in the economic development which derives from technical change.

The internationalization of R&D represented by the technical enterprise can be a benefit to nations able to form linkages with it. Nations whose policies might prevent those linkages may be induced to change those policies in order to gain the benefits. This applies, though in different ways, for both industrialized and developing countries.

Finally, this internationalization relates the health of the technical enterprise to the political health of the nations which cooperate in its unrestricted operation. Economic development and growth requires a strong technical base. This depends at the least on access to large markets and, even then, on the ability to draw upon the technical enterprise to participate in technical advances within high technology areas. Both call for internationalization of trade and technology. Any nation that withdraws from that system, whether for economic or political reasons, weakens its technical base with respect to others that work within the technical enterprise. Thus, a nation wishing to maintain a strong, high-level base of modern technology must develop working relationships internationally that make it part of the networks that constitute the technical enterprise. This, of course, is consistent with an outward-looking growing economy, which is a vital element in the political health of any country. The three critical areas—technical base, economic development, political stability—are interdependent in the modern industrialized world. The technical enterprise is both the symbol and the practical implementation of that interdependence.

NOTES TO CHAPTER 14

1. H.I. Fusfeld and R.N. Langlois (eds.), *Understanding R&D Productivity* (New York: Pergamon Press, 1982).
2. Carmela S. Haklisch, "Technical Alliances in the Semiconductor Industry," Center for Science and Technology Policy, Graduate School of Business Administration, New York University, 1986.

INDEX

305

ABOUT THE AUTHOR

Herbert I. Fusfeld has enjoyed a rich exposure to the principal features of the American technical system, having spent more than thirty years as a research manager. For the first five of those years he was head of the physics and mathematics division in a government defense laboratory (Frankford Arsenal). He then served for twenty-five years as director of research for two Fortune 500 corporations, AMF and Kennecott Copper. While at AMF, he planned and established a research laboratory in England, learning first-hand the conditions and attitudes of the European industrial research community.

Dr. Fusfeld served as president of the Industrial Research Institute (1973–74) and was appointed to the U.S.–U.S.S.R. Joint-Commission for Scientific and Technological Cooperation, the Advisory Council of the National Science Foundation, and the State Department Advisory Committee on International Investment, Technology and Development, and many other organizations.

In 1978 he established the Center for Science and Technology Policy currently at the Graduate School of Business Administration, New York University. As director of the Center, he supervises and is engaged in study of the process of technical change.